CAMBRIDGE LIBRARY COLLECTION

Books of enduring scholarly value

Cambridge

The city of Cambridge received its royal charter in 1201, having already been home to Britons, Romans and Anglo-Saxons for many centuries. Cambridge University was founded soon afterwards and celebrates its octocentenary in 2009. This series explores the history and influence of Cambridge as a centre of science, learning, and discovery, its contributions to national and global politics and culture, and its inevitable controversies and scandals.

The Shaping of Cambridge Botany

Originally published in 1981, this volume marked the 150th anniversary of the acquisition by the University of Cambridge of the site for its 'New Botanic Garden'. Written by a distinguished authority on British and European plants, the then Director of the University Botanic Gardens, the book honours the eminent scientists and key ideas that have been most influential not only in the history of the Botanic Gardens but also in guiding the development of botany itself from the foundations laid by John Ray in the mid-seventeenth century. It also includes rarely seen archival material . The core theme of the book is whole-plant botany, as distinct from cell biology or the study of the 'lower plants' (bacteria and fungi). Relatively little emphasis is given to genetics, plant physiology or ecology. The reader is nevertheless richly rewarded by this engaging and erudite account of Cambridge botany over more than three centuries.

Cambridge University Press has long been a pioneer in the reissuing of out-of-print titles from its own backlist, producing digital reprints of books that are still sought after by scholars and students but could not be reprinted economically using traditional technology. The Cambridge Library Collection extends this activity to a wider range of books which are still of importance to researchers and professionals, either for the source material they contain, or as landmarks in the history of their academic discipline.

Drawing from the world-renowned collections in the Cambridge University Library, and guided by the advice of experts in each subject area, Cambridge University Press is using state-of-the-art scanning machines in its own Printing House to capture the content of each book selected for inclusion. The files are processed to give a consistently clear, crisp image, and the books finished to the high quality standard for which the Press is recognised around the world. The latest print-on-demand technology ensures that the books will remain available indefinitely, and that orders for single or multiple copies can quickly be supplied.

The Cambridge Library Collection will bring back to life books of enduring scholarly value (including out-of-copyright works originally issued by other publishers) across a wide range of disciplines in the humanities and social sciences and in science and technology.

The Shaping of
Cambridge Botany

S. M. WALTERS

CAMBRIDGE
UNIVERSITY PRESS

CAMBRIDGE UNIVERSITY PRESS

Cambridge New York Melbourne Madrid Cape Town Singapore São Paolo Delhi

Published in the United States of America by Cambridge University Press, New York

www.cambridge.org
Information on this title: www.cambridge.org/9781108002301

© in this compilation Cambridge University Press 2009

This edition first published 1981
This digitally printed version 2009

ISBN 978-1-108-00230-1

The Shaping of Cambridge Botany

Frontispiece: *Rosa* 'Cantabrigiensis', a spontaneous hybrid studied by Hurst in the Botanic Garden, and given a Royal Horticultural Society Award of Merit in 1931.

Rosa 'Cantabrigiensis'

The Shaping of
Cambridge Botany

A short history of whole-plant botany in Cambridge
from the time of Ray into the present century

by
S. M. WALTERS
Director of the University Botanic Garden, and
Fellow of King's College, Cambridge

Published on the occasion of the sesquicentenary of
Henslow's New Botanic Garden, 1831–1981

Cambridge University Press
Cambridge
London   New York   New Rochelle
Melbourne   Sydney

Published by the Press Syndicate of the University of Cambridge
The Pitt Building, Trumpington Street, Cambridge CB2 1RP
32 East 57th Street, New York, NY 10022, USA
296 Beaconsfield Parade, Middle Park, Melbourne 3206, Australia

First published 1981

Phototypeset in Linotron 202 Bembo by
Western Printing Services Ltd, Bristol

Printed in Great Britain at the
University Press, Cambridge

British Library Cataloguing in Publication Data

Walters, Stuart Max
    The shaping of Cambridge botany.
    1. University of Cambridge. School of Botany – History
    I. Title
    581'.07'1142659   QK51.2.G7   80–41204

ISBN 0 521 23795 5

Contents

	List of illustrations	vii
	Preface and acknowledgements	xi
1	Introduction: botany, medicine and horticulture	1
2	Ray and the herborising tradition	6
3	Bradley and the horticultural tradition	15
4	The Martyns and the Linnaean tradition	30
5	Henslow and the rise of natural science	47
6	Babington, Vines and Lynch: the fragmentation of botany	65
7	The New Botany School: Marshall Ward and his successors	83
8	Whole-plant botany and the modern Botanic Garden	95
	Bibliography	111
	Index	117

'It seems, as one becomes older,
That the past has another pattern, and ceases to be a mere sequence –
Or even development: the latter a partial fallacy
Encouraged by superficial notions of evolution,
Which becomes, in the popular mind, a means of disowning the
    past.'

T. S. Eliot: *Four Quartets: The Dry Salvages*

. . . We would urge men of University standing to spare a brief
interval from other pursuits for the study of nature and of the vast
library of creation so that they can gain wisdom in it at first hand
and learn to read the leaves of plants and the characters impressed on
flowers and seeds. Surely we can admit that even if, as things are,
such studies do not greatly conduce to wealth or human favour,
there is for a free man no occupation more worthy and delightful
than to contemplate the beauteous works of nature and honour the
infinite wisdom and goodness of God. . . . Of course there are
people entirely indifferent to the sight of flowers or of meadows in
spring, or if not indifferent at least pre-occupied elsewhere. They
devote themselves to ball-games, to drinking, gambling, money-
making, popularity-hunting. For these our subject is meaningless.
We offer a hundred banquets to . . . the true philosophers whose
concern is not so much to know what authors think as to gaze with
their own eyes on the nature of things and to listen with their own
ears to her voice; who prefer quality to quantity, and usefulness to
pretension: to their use, in accordance with God's glory, we
dedicate this little book and all our studies.

John Ray: Preface to *Catalogus*
Plantarum circa Cantabrigiam nascentium,
1660 (trans. Raven)

Now we have taken this fhort View of Nature and its Order, we
may judge how fhocking and deteftable muft every Thing be, that
is contrary to it; its Beauty is Freedom, and its Gaiety familiar . . .
Nature is full of Variety, and it is the great Variety in Nature that
captivates the Mind, and draws Admiration.

Richard Bradley in
General Treatise of Husbandry & Gardening
(the first horticultural periodical),
Aug.—Sept. 1724, 14; from
Collected Writings on Succulent Plants, Ed. G. Rowley

List of illustrations

Frontispiece: *Rosa* 'Cantabrigiensis'
1   *Geranium sanguineum* and the Devil's Dyke.
2   John Ray.
3   Pages from John Martyn's copy of Ray's *Catalogus*.
4   Loggan's Plan of the Oxford Botanic Garden.
5   The three species of *Primula*.
6   Frontispiece from Ray's *Methodus*.
7   Drawings of seeds from the essay on seeds in the *Methodus*.
8   Specimen of the Pasque Flower, *Pulsatilla vulgaris* Miller, from Ray's own Herbarium.
9   A page of the *Catalogus Plantarum Angliae*.
10   Specimen of the cornfield weed *Silene gallica*.
11   Coffee plant.
12   Map used by Stephen Hales and his friends.
13   An account of the Ananas or West Indian Pine Apple.
14   Monthly notes for November 1724, from Bradley's *General Treatise*.
15   Title-page and illustration from Bradley's *The Virtue and Use of Coffee*, 1721.
16   Illustration from Bradley's *Philosophical Account of the Works of Nature*.
17   Title page and frontispiece from Bradley's *New Improvements of Planting and Gardening*.
18   Description and illustration of the Tulip-tree, *Liriodendron tulipifera*.
19   Title-page and illustrations from a manuscript by Bradley.
20   Title-page and frontispiece from Bradley's *Country Gentleman and Farmer's Monthly Director*.
21   Title-page and frontispiece from Bradley's *The Riches of a Hop-Garden Explained*, 1729.
22   Illustration of pollination.
23   *Martynia* and the gates of the Chelsea Physic Garden.
24   Illustrations of floral structure from John Martyn's *First Lecture of a Course of Botany*, 1729.
25   Unpublished illustration of *Pilularia* by John Martyn.
26   Preface to Martyn's *Historia Plantarum Rariorum*, 1728.
27   Illustration of a *Pelargonium*.
28   Fruit of *Martynia* (*Proboscidea*).
29   Illustration of *Passiflora* from Martyn's *Historia*.
30   Specimen in Martyn's Herbarium.

31   Title-page and frontispiece of John Martyn's copy of Linnaeus' *Flora Lapponica*, 1737.

32   Page from Thomas Martyn's *Plantae Cantabrigienses*, 1763.

33   Dedication in Thomas Martyn's translation of Rousseau's *Letters on the Elements of Botany*, 1785.

34   Illustration of *Viola* from Thomas Martyn's *Thirty-eight Plates* . . ., 1788.

35   Heberden's *Course of Lectures*, 1747.

36   View of the main entrance to the Old Botanic Garden.

37   Ackermann print of The Old Botanic Garden.

38   Herbarium specimen of *Silene maritima* made by Donn.

39   *Dipsacus strigosus* in Little St Mary's Churchyard.

40   Lecture-rooms in the Old Botanic Garden, as Henslow knew them.

41   Illustrations from Henslow's early paper on the geology of Anglesey, 1821.

42   Two of Henslow's 'Botanical Diagrams'.

43   Henslow's specimen of Box leaves.

44   Plan of the Old Botanic Garden.

45   Site of the present Botanic Garden in 1809.

46   Old label on Commemorative Lime Tree at entrance to Garden.

47   Lapidge's design for the glasshouse range dated 1830.

48   Lapidge's design for the 'New Botanic Garden' dated 1830.

49   Plate from Henslow's booklet prepared for the day excursion from Hitcham to Cambridge on 27 July 1854.

50   Portrait of Henslow by Maguire, 1849.

51   Bridge of Sighs, St John's College, with *Elodea canadensis*.

52   Yew Tree (*Taxus baccata var. dovastoniana*), planted by Babington in 1843.

53   Staff of the Botanic Garden in 1876.

54   Allotment advertisement from the time of Lynch.

55   *Opuntia cantabrigiensis*.

56   Staff of the Botanic Garden in 1912.

57   *Thladiantha dubia*.

58   The glasshouse range in 1911.

59   *Lathraea clandestina*.

60   An Edwardian postcard.

61   The main gates on Trumpington Road.

62   Lime Tree (*Tilia × europaea*).

63   The old *Sophora* tree on the Old Botanic Garden site, 1931.

64   Marshall Ward's Elementary Botany class in 1906.

65   *Ginkgo biloba*.

66   Cover of the third number of the *Tea Phytologist*.

67   Illustration by Vulliamy from Seward's *Plant Life Through the Ages*.

68   Frontispiece from Elwes and Henry's book on trees.

69   *Platanus cantabrigiensis* Henry.

70   Plate from Hurst's account of the genetic history of Roses.

71   The dwarf Pine, *Pinus sylvestris* 'Moseri', on the Old Rock Garden, *c.* 1927.

72   Willows by the River Cam.

73   Humphrey Gilbert-Carter.

74   Sunday tea-party at Cory Lodge.
75   Student excursion at Swaffham Prior, 1942.
76   *Magnolia* plate from Hickey and King 1980, drawn from live material
      grown in the Garden.
77   Old entrance to the Garden from Bateman Street, *c.* 1935.
78   Cedar of Lebanon, *Cedrus libani.*
79   Vertical aerial photograph of the Garden taken on 26 August 1945.
80   *Cytisus battandieri.*
81   View of the Garden in 1953.
82   Staff of the Garden in 1959.
83   The limestone mound in the Ecological Area.
84   Conservation display in the Alpine House, 1979.

The illustrations at the ends of the chapters are reproduced from works by
Richard Bradley.

To Lorna

Preface and acknowledgements

An institutional anniversary provides the occasion to stand back from the transient pre-occupations of administration, teaching and research, and look at the tradition in which the particular institution operates. In my own case, several factors have coincided to make the process especially congenial. The first, and most important, is very personal, and concerns my attitude to history. Twenty years ago, the suggestion that I should write a book on the history of the Botanic Garden, or of Cambridge botany, would have worried and even depressed me: now I find the opportunity richly rewarding. I can only report this change of heart without comment.

A second factor is the product of my career as Curator of the Herbarium and Lecturer in the Botany School, from which I was appointed in 1973 as Director of the Garden. This translation from a professional career in scientific botany in the main University Department to my present post enables me to appreciate the separate and combined elements in two interestingly different traditions, and stimulates me to ask how the differences came about. To some extent, this book is a product of such questioning.

The third of the factors encouraging me to write this little book concerns the nature and size of the University Botanic Garden itself. An institution occupying under 40 acres and employing fewer than 40 people is a comprehensible whole, in which it is possible to feel personal links and loyalties and to understand the nature and strength of tradition. I am peculiarly fortunate in that I lived, for the whole 25 years of my Botany School career, in the corner of the Garden, so that the beauty and value of the collections were part of my own background, and indeed that of my wife and family also. We took it for granted that a benevolent university should provide such a gracious mixture of science and amenity, and I gradually came to recognise who or what was responsible for this extraordinarily enlightened policy. I hope the book will make clear what I now see as the important figures and ideas in the history of the Garden and Cambridge botany, and the part played by certain men in shaping that history.

The fourth element in the picture is the contribution of my predecessors. My debt to Humphrey Gilbert-Carter, first Director of the Garden, will become apparent to readers in the main text of the book, and needs no further comment. I owe, however, a debt of similar importance to my immediate predecessor, John Gilmour, for two influences in particular: for his infectious enthusiasm for the theory and philosophy of classification, and especially of biological taxonomy, which led me to read the history of my subject with care; and more concretely for his work in creating and furnishing, over the first twenty years of the life of the Cory Fund, the excellent horticultural and botanical Library which the Garden now possesses. Without reference works to hand, such as the *Dictionary of National Biography*, or the recent study of *British Botanical and Horticultural Literature before 1800* by Blanche Henrey (a book, incidentally, which owes its very existence to John's bibliographic interests), my task would have been incomparably more difficult. F. G. Preston, Superintendent of the Garden in my student days, had published in 1940 a paper on the history of the Garden which stimulated my interest and provided a basis on which to build. In explaining my debt to the third of my teachers, Professor Sir Harry Godwin, I find a special difficulty, for he has guided and influenced not only my academic career but even, with his characteristic kindness and generosity, the detailed shape of this book. The allusions in the main text to the value and importance of Sir Harry's influence I have consciously left in a very personal form, hoping that if there is merit in my career and my writing, credit may go to the right place.

A fifth factor, which obviously overlaps, is the helpful and encouraging attitude of my contemporaries. I succeeded to the Directorship of the Garden when the late Professor Brian was in the Chair of Botany, and benefited greatly from his wise, tolerant guidance at a time of some difficulty, when the role and importance of the Garden were, reasonably enough, under close scrutiny because of financial stringency. In particular I recall that he was very encouraging when I first suggested a possible celebration of this 150th Anniversary, and we discussed the publications, including this book, which might be associated with it. Though Percy Brian's research fields were in biochemistry and mycology, he had a genuine interest in whole-plant botany, strengthened by a passion for gardening; for him the Botanic Garden tradition needed no laborious explanation or defence. From his successor, Richard West, who shares with me and many others the distinction of being a pupil of Godwin, I have naturally received full encouragement, and Richard is responsible for the suggestion, made early in our discussion of the possible scope of the present book, that I might enlarge it so that it tells a story of Cambridge botany over more than three centuries.

Writing such a book as this *in medias res* brings its problems, of which the most obvious is shortage of time. Many things could have been more completely or more felicitously explained had I been able to drop everything else and pursue every hare which started up. Perhaps the imperfections and inadequacies are most evident in the later chapters, where the relevant material is so plentiful that I cannot but be aware of my neglect of important sources. That these inadequacies are not even more crude and glaring is in great measure due to several colleagues who have very kindly read and criticised part or all of the draft text. In particular I must thank, in addition to those already mentioned, my colleague David Coombe, whose detailed knowledge of the history of Cambridge botany has been put very generously at my disposal, and Professor David Webb (Trinity College, Dublin) and Mr Arthur Chater (Department of Botany, British Museum, Natural History), both Cambridge men, who have kindly read and criticised my writing in various stages.

In using the title I have chosen, I am aware that the contents of my book may be a disappointment to some readers. A tour of the Botany School in Cambridge today reveals how diversified the subject has become and how, in particular, several areas of development in the present century make much botanical research quite different in kind from what went on before. Two of these need special comment. The first is epitomised by the rise of the term 'cell biology', and its implied contrast with the 'whole organism' biology with deep traditional roots. The sub-title I have chosen is intended to emphasise that it is the science of the whole plant which is my theme. The second is the study of lower plants, and in particular the rise, mainly in the present century, of that part of biology concerned with the diseases caused by fungi and bacteria. This is no part of my story, however important it may be to modern agriculture, forestry and medicine. A more serious misunderstanding may, however, arise with regard to those 'whole-plant' subjects which I have not explicitly excluded from consideration, but for which my account is ludicrously inadequate as a review of the history of any of those subjects themselves. I refer especially to genetics, plant physiology, and ecology. An account of Cambridge genetics which does not mention Catcheside, Fisher, Whitehouse or Thoday, or Cambridge plant physiology which says nothing of the work of F. F. Blackman or G. E. Briggs, or Cambridge ecology which ignores the contribution of A. S. Watt, would be a travesty indeed. My excuse for such extraordinarily cavalier treatment is that I am concerned with the factors shaping the 'new botany' of the present century, rather than the recent history or, even less, what goes on in the Department at the present day. In setting this as my goal, I run the risk of offending those whose work receives no mention; my only defence is that I had to define some limits, and try to hold to them.

I have reserved a special position to record my indebtedness to Dr Raymond Williamson, of Clare College, who has a unique knowledge of Richard Bradley and an almost complete set of Bradley's works. Not only has Dr Williamson kindly made available to me a mass of excellently-ordered unpublished notes on Bradley's writings, which saved me a great deal of time-consuming search, but he has confirmed his intention to bequeath his set of Bradley's books to the Botany School, where it will greatly improve our remarkably inadequate representation of the published works of the first Professor of Botany.

I am grateful to the Cambridge University Press as publishers for the decision to provide lavish illustration, and for all the help I have received in the pleasant task of assembling the illustrations. A special word of thanks is also due to Geoff Green, who designed the book. The head-pieces for the Chapters and other plant illustrations have been specially drawn by Michael Hickey, a former student of the Garden, and a grant towards the expenses of preparing these illustrations has been made by the Botanic Garden Association, CUBGA. It is also a pleasure to record my indebtedness to Graham Thomas, Gardens Consultant to the National Trust, and a former student of the Garden, for the water-colour illustration of *Rosa* 'Cantabrigiensis' which makes an excellent frontispiece. The decision to include a colour frontispiece was taken late in the planning of the book, and it was particularly good of Mr Thomas to undertake the painting, based on colour photographs we provided, and sketches by Mrs D. M. Watson who worked with flowering material from the Garden. The cost of printing this colour frontispiece has been borne by the Cory Fund.

In conclusion I must record my special thanks to members of the Botanic Garden staff and others who have helped in important ways during the preparation of this book. My thanks go first to my colleagues Peter Orriss, Superintendent, and Peter Yeo, Taxonomist and Librarian, from both of whom I have had encouragement and much practical help. For special help with bibliographic matters I would wish to thank particularly the Assistant Taxonomist and Librarian, Clive King; not only has he carefully checked the references, but he has also in recent years helped me to arrange the sorting and transfer of much of our historic archival material to the University Library. To the Assistant Archivist in the University Library, Dr Elizabeth Leedham-Green, I am grateful for her efficient cataloguing of all the Garden archives, including those retained by us; and I must also record my thanks to Dr Christine

Quartley, who has in recent years given excellent voluntary service to the Garden in sorting and indexing our correspondence. I am especially grateful to Mrs Betty Attmore and Mrs Anne James for taking my badly-written manuscript and converting it so quickly and correctly into a typescript suitable for publication. My final thanks I reserve for my wife; not only has she compiled the index, but she has also borne with characteristic loving care the brunt of my sometimes unreasonable enthusiasm to finish this book.

S. M. Walters
March 1980

We wish to thank the following owners and copyright holders for permission to reproduce their material. The British Museum (Natural History): illustrations 8, 10; Cambridge Evening News: illustration 81; Cambridge University Botany School: illustrations 3, 6, 7, 19, 20, 21, 25, 26, 27, 29, 30, 31, 32, 33, 34, 38, 43, 66; Cambridge University Library: illustrations 37, 42, 47, 48, 49; Cambridgeshire Collection, Cambridgeshire Libraries; illustrations 44, 60; Cory Library, Cambridge University Botanic Garden: illustrations 13, 14, 16, 17, 18, 22, 24, 68; The Master and Fellows of St John's College, Cambridge: illustration 50; The Master and Fellows of Trinity College, Cambridge: illustration 15; Mr John Thirsk: illustrations 46, 78; Mr G. S. Thomas; illustration 71.

I

Introduction: botany, medicine and horticulture

There are many books on the history of biology, or of botany as a separate science, but it is not easy to distil from them certain relevant ideas which I believe to be important in understanding the shape of modern botanical science. Indeed, I have been forced to the conclusion that much of the written account of the history of biology is liable to be misinterpreted by many readers. Since I was myself until recently guilty of such misinterpretation, I feel that it would be useful to clarify the matter here, before we start on the particular history of the Cambridge School of Botany and the Cambridge Botanic Garden.

It is common knowledge that the science of botany developed from a study of plants as useful to cure diseases, and that the oldest surviving Botanic Gardens, at Padua and Pisa in Italy, were founded in the middle of the sixteenth century as 'herb gardens' whose primary purpose was to grow the medicinal plants important to the medical science of the time. The foundation of such gardens in connection with universities and centres where there were flourishing medical schools proceeded quite rapidly in the late sixteenth and early seventeenth centuries, and the first British Botanic Garden was founded in 1621 in the University of Oxford. The background to seventeenth- and eighteenth-century botany can be found in the many 'Herbals' or illustrated books devoted to the identification and description of medically useful plants, a tradition which occupied the late medieval and Renaissance periods in Europe, and stamped the Linnaean botany which succeeded it with an individual shape which persists in the botanical taxonomy of the present day. This medical background to botany is well described by Arber (1912); its effect in shaping the modern classification of the flowering plants I have myself discussed (Walters 1961).

Whilst the origins of botany in medicine are well described and documented, there is surprisingly little discussion in the standard works of the reasons why botany developed in this way. It is, of course, true that a chronological sequence of events in history may be known and relatively undisputed, whilst the interpretation of the same history can remain obscure and controversial. The difficulty enshrined in the aphorism *'post hoc, ergo propter hoc'* is obvious enough, in history as in experimental science. Yet a medical origin for European botany is not self-evidently necessary, and the sequence of events does demand some explanation. In particular, it seems reasonable to ask why agriculture and horticulture, both ancient and very important 'sciences' dealing with plants, contributed little or nothing to modern botany (or indeed to biology as a whole) until relatively recent times. Bound up with this question are two larger questions, which take the enquiry further back in history. One concerns the origins of biology as a recognisable, separate science, and its relation to botany and zoology. The other has even wider ramifications, and concerns the difference between observational and experimental science.

Let us examine a little more closely the relation between biology and medicine. As we see it now, in terms of twentieth-century science, medicine might look like a peculiar sort of applied biology. Man is an animal, and animals, directly or indirectly, are dependent on plants for their food. Indeed, there is no logical reason why today we should not go to a Department of Applied Biology to seek 'medical' advice. But this is not the structure we have, and the reason is embedded in history. To understand the shape of science in any modern university we must go back to the beginnings of universities in Europe in the twelfth and thirteenth centuries. As Green (1969) explains in his very readable survey of the history of British universities, in the twelfth century 'monastic and cathedral schools were already in existence but . . . the monasteries . . . were no longer the centres of culture they had once been . . . Those who wanted to study and to teach gathered together in groups independent of the monastic and cathedral schools and, living as often as not in a strange city, found it necessary to create an organisation, a guild, a *universitas societas magistrorum discipulorumque*, to safeguard their position in the community, to defend their privileges and to order their own lives. It was in this way that universities came into existence and teaching became institutionalised.' Green continues: 'the rapid growth of the medieval universities testified to the social and intellectual need which they were meeting. They brought into existence an entirely new class of educated men, academics and intellectuals . . . If in the first instance the universities trained a clerical caste which dominated the culture and government of western Europe, ultimately they helped also to instruct the laymen, lawyers and doctors in particular, who were to challenge ecclesiastical control.'

Against this background we can imagine a medieval university scholar with an interest in what we would now call 'natural history'. The formal courses of study open to him were in theology, law and medicine. Only in medicine could he pursue his interest, and even there only, effectively, in botany. Thus botany grew as an answer to the question: what are all these different plants growing wild, and which of them are useful to cure diseases and injuries? It is significant that zoology did *not* grow in answer to a parallel question. Medicine, in fact, asked a very different question about animals, which could be put as follows: how closely does man resemble the other animals, and is there, then, a hierarchy of animals, with Man at its head, a hierarchy which culminates in Man? This is the familiar Aristotelian '*scala naturae*', a very potent force in zoological thought which has no traditional counterpart in botany.

It is for these reasons misleading to think of 'biology' as having originated from medicine. Certainly both botany and zoology had such origins, but they were separate, and the origin of botany as a science is very much older than zoology. This is very clearly illustrated by the history of the two Cambridge Chairs; as we shall see, the first Professor of Botany was appointed in 1725, whilst the university had to wait until 1869 for the zoological chair, which was eventually founded in that year as the 'Chair of Zoology and Comparative Anatomy'. Over-simplifying greatly, one could sum up the difference by saying that, in the late medieval university, you could, if you were a naturalist, study something recognisably the science of botany, but any 'zoological' study would be more or less indistinguishable from medicine, and would be dominated by human anatomy and physiology.

We are now equipped to give a tentative answer to the question about the exclusively medical origins of botany. Theoretically, an applied science of botany could have arisen from man's total experience of using plants, not simply from their known or presumed medicinal uses. Indeed, if we study the languages and cultures of primitive tribes, we find quite elaborate systems of botanical naming and classification ('folk taxonomies') which *do* cover the total value of the plant world to man – food, poison, shelter, religious ceremony, etc – and not simply medicine. Why did medicine dominate?

The answer seems to lie in the class structure of medieval society. The Church, law and medicine were the three great professions for which the medieval university trained its students. To that extent the ancient universities were more akin to modern polytechnics, training men to fit the available careers in the society of the time, than to the 'liberal University' of the eighteenth century which pursued learning for its own sake. [It is, perhaps, the strength of Oxbridge that it retains the tradition of vocational training side by side with the liberal, academic tradition: the tension between the two is particularly fruitful in the natural sciences.]

With these considerations in mind, it is extraordinarily interesting to turn to Theophrastus, botanical pupil of Aristotle, and see what plant science contained, as known to the ancient Greeks in the fourth century B.C. The *Enquiry into Plants* (trans. Hort 1916) consists of nine books of 'applied botany', covering all wild and garden plants and their use by man. The surprise is that only a part of *one* of these nine books, the last in fact, is devoted to the medicinal value of plants. The Aristotelian School was quite free from medical bias and dominance, and any botanical science directly descended from Theophrastus without distortion would have incorporated experimental agriculture and horticulture in its tradition without question. Medical dominance, visible in the writings of Dioscorides in the first century A.D., far from 'stimulating' the development of botany, might be thought on this analysis to have captured and stultified it. In particular, it favoured what we might now call botanical taxonomy and discouraged other, equally legitimate, forms of enquiry. We shall see some of the interesting implications of this as they are worked out in the history of Cambridge botany.

How does this fundamental difference between the medical origins of botany and zoology relate to the difference between 'observational' and 'experimental' science? Here we have one of the most interesting, and at the same time most complex, of issues which are relevant to our present thesis. Our starting-point is the demand from medicine that the botanist (or 'herbalist') should accurately describe and identify *all* plants, so that their medicinal properties can be assessed. Because medicine had (and still has) a privileged status in society as a whole and in the institutions of higher learning in particular, it was always respectable for naturalists to herborise and know their plants in the field. No such continuous tradition is present in field zoology. Indeed there is no term 'to zoologise', equivalent to the term 'to botanise'. The sixth Professor of Botany in Cambridge, Charles Cardale Babington, published in 1843 a *Manual of British Botany* which is, purely and simply, a Flora of the British Isles, a technical work describing and identifying all higher plants found in Britain. Such a narrow interpretation could not operate in university Zoology. As we have seen, there *was* in fact no Professor of Zoology in Cambridge when Babington published his *Manual*, and the subject was still a part of Comparative Anatomy. A zoology born of comparative anatomy and linked to, even dominated by, studies of human anatomy and physiology, was much more naturally inclined to consider together form and function, observation and experiment, than was the traditional observational botany.

There is another aspect of the difference between the two traditions which has had a great effect on the way botanists and zoologists think and write today. This is the Aristotelian '*scala naturae*', an evolutionary picture which fixed a zoological hierarchy with man as the highest animal at the head. No corresponding idea is fixed in pre-Darwinian botany, and botanists even today are free to be sceptical of both the broad pictures of the course of organic evolution and of the more rigid Darwinian orthodoxies which attempt to explain them. Both these themes – the rise of experimental science, and the impact of Darwinism – find their place in the closing chapters of the book.

Figure 1 *Geranium sanguineum* and the Devil's Dyke where John Ray first recorded it in 1660.

2
Ray and the herborising tradition

A history of Cambridge botany must necessarily be selective, especially when it deals with those more distant periods for which it is partly an accident of historical research which figures seem of special importance. There can, however, be no doubt about the contribution of one man in establishing the continuous botanical tradition which survives and even flourishes to the present day. That man is John Ray, son of an Essex blacksmith, who was admitted a Sizar in Catherine Hall (now St Catherine's College) in 1644 at the age of 17 years, and transferred to Trinity College in 1646, where he was elected a Fellow in 1649. There he held a selection of teaching and administrative posts until the year 1662, when, like many of his colleagues, he refused assent to the Act of Uniformity and was forced to resign his College appointments and leave the university.

The excellent, scholarly biography of John Ray written by Charles Raven in 1942 (2nd edn 1950) and the detailed bibliography by Keynes (1951) provide abundant material from which to assess the character and scientific ability of this very remarkable man. It would be inappropriate in a limited space to attempt even a condensed general account of his achievements, but there are several aspects of Ray's life and work which are highly relevant to the special contribution he made to botany, and Cambridge botany in particular, and which we should at least briefly mention.

To understand Ray's contribution, we need to see the background of culture and factual knowledge available to him. The explanation in the Preface to the 'Cambridge Catalogue' – his Cambridgeshire Flora, published in 1660, which we describe below – as to how he came to write the book is really so good that we should let it speak for itself (original in Latin; trans. Ewen & Prime 1975).

I became inspired with a passion for Botany, and I conceived a burning desire to become proficient in that study, from which I promised myself much innocent pleasure to soothe my solitude. I searched through the University, looking everywhere for someone to

Figure 2 John Ray: an engraving by G. Vertue, 1713, from a portrait by
W. Faithorne.

act as my teacher and my guide, who would instruct me and, so to
speak, initiate me, so that I would be able to enjoy the benefit of his
advice whenever I needed it. But, to my astonishment, among so
many masters of learning and luminaries of letters I found not a single
person who was deeply versed in Botany, and only one or two who
had even a slight acquaintance with the subject. . . . What was I to do
in this situation? Should I allow the flame of my enthusiasm to be

quenched or diverted to some other field of study? I decided that this must not happen . . . Why should not I, endowed with ample leisure, if not with great ability, try to remedy this deficiency so far as my power permitted, and advance the study of Phytology, which had been passed over and neglected by other men?

So far as the University was concerned, it is clear that Ray could get little help. Indeed, it becomes apparent when one inspects the copious bibliography ('*Explicatis*') attached to the 'Cambridge Catalogue' that Ray found relatively little help from any previous botanical work published in England. There were in fact only four important English botanical authors available to him – two from the previous century and two from nearer his own time. The first of these chronologically was a Cambridge man, William Turner, often called 'the Father of British Botany', who studied medicine at Pembroke College, Cambridge and published in 1551 the first part of his *New Herball*. Ray obviously found this book useful, and commends Turner as 'a man of sound learning and judgment'. In the words of John Gilmour (1944), Turner's *New Herball* provided for the British medical profession 'for the first time . . . a volume in their own tongue which gave them some reasonably accurate information on the plants available for their work'.

Ray had far less respect for the other famous botanical writer in sixteenth-century England, John Gerard, whose *Herbal* dates from 1597. According to Raven (1950, p.74), 'the charm and interest of the book are apt to blind its readers to its defects, the attachment of plates to the wrong descriptions, the reckless multiplication of species, the credulity and errors, the false claims and statements, and the blunders due to ignorance of Latin'. Ray realised these defects in the original work, and used the *Herbal* in the much improved edition made by Thomas Johnson in 1633, speaking appreciatively of Johnson's worth as a botanist.

The difference between Gerard and Johnson, both practising medicine in London, is most obvious in their attitude to wild plants. To Gerard, the successful surgeon, the many plants he knew and cultivated in his London garden were of interest only for their medicinal value. Johnson, on the other hand, combined his profession as an apothecary with a real enjoyment of field botany, and might be said, together with his London friends, to have created the tradition of herborising which Ray so successfully continued. Johnson's *Iter* and *Descriptio*, published in 1629 and 1632, are choice botanical classics now available in translation (ed. Gilmour 1972).

Before we leave Gerard, we should record a curious piece of evidence, preserved in the Lansdowne manuscripts in the British Museum, about the earliest attempt to found a Garden in

Cambridge. This is a letter, drafted by Gerard in 1588 for Lord Burghley, then Chancellor of the University 'to signe for ye University of Cambridge for planting of gardens'. It commends Gerard himself to the University as a suitable person to look after their Botanic Garden 'by reason of his travaile into farre countries his great practise and long experience'. There is no evidence that the University ever received such a letter, and the effort came to nothing.

The most modern work available to Ray in English was by John Parkinson, whose *Theatrum Botanicum* was published in 1640. It is, according to Raven (1947), 'an old man's book, diffuse and often ungrammatical in its long and ill-penned sentences', but nevertheless a work revealing 'the authentic passion for a garden and the quiet wisdom of a gardener, than which there are a few things more precious'. His earlier work with the punning title *Paradisi in sole Paradisus terrestris* (The Park on earth of the Park in sun), published in 1629, has a good claim to be the first English gardening book: it represents an important and already separate tradition of plant lore which increasingly diverged from the medically-dominated academic botany, as we shall see in later chapters.

With this background, we can now look at the nature of Ray's contribution to English botany in the 'Cambridge Catalogue'. It is a small volume entitled *Catalogus plantarum circa Cantabrigiam nascentium*, published in Cambridge by John Field, University Printer. The text is in Latin throughout, except where Ray gives a Cambridgeshire locality, when the sentence or phrase is given in italics in English. These English phrases have delighted generations of Cambridge botanists, to whom until recently they were the only immediately comprehensible parts of the text. Now we have the very useful translation by Ewen & Prime (1975) which makes Ray's Flora accessible to all.

Ray's purpose in writing the book is clear from the quotation already given from his Preface. He was anxious to do something to promote a proper study of Botany in the University, and had found from his own experience that herborising with friends was an excellent way of combining scientific study with a healthy outdoor pursuit. He hoped that what he had learned would help others coming after him 'so that they, perhaps less patient of labour than myself, would not be deterred by the endless succession of difficulties, and falter in their studies'.

Bound up with this very laudable aim to share his enthusiasm with others of like interest, there was undoubtedly also a very real desire on Ray's part to rescue botany from a neglect which was the more serious because other branches of 'natural philosophy . . . were flourishing and making progress in our midst'. The Royal

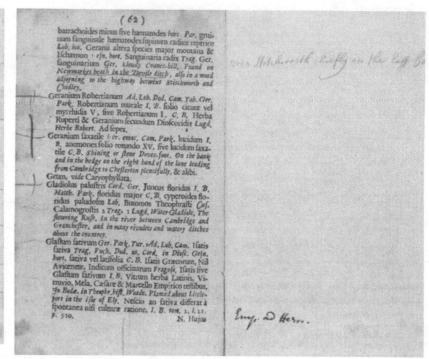

Figure 3 Pages from John Martyn's interleaved and annotated copy of Ray's *Catalogus*, with the reference to *Geranium sanguineum*.

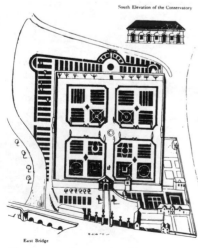

Figure 4 Loggan's Plan of the Oxford Botanic Garden in 1675.

Society was founded in 1662, and Ray was to see in his own lifetime a remarkable burst of what we would now call 'natural science'. Ray himself was elected a Fellow of the Royal Society in 1667, and Isaac Newton began his career as a young student at Trinity during Ray's time there as a Fellow. Part of the neglect which concerned Ray was due to the absence of any formal teaching of botany – or any other science if medicine be excepted – in the University. During Ray's time at Trinity there were in fact only seven Professors in Cambridge: two of Divinity, and one each of Arabic, Greek, Hebrew, Civil Law and Medicine.

Oxford, by contrast, was well provided for; the Botanic Garden there was founded in 1621, the earliest in Britain, and Robert Morison was appointed Professor of Botany in 1669. There is no evidence that Ray visited the Oxford Botanic Garden before 1669, but in May of that year he was there, and, according to Raven (1950, p. 150) 'impressed by the work of Jacob Bobart junior and William Browne of Magdalen College, . . . at the Botanic Garden, of which Bobart had just succeeded his father as custodian'. Apparently the young Bobart told Ray on this occasion that the progeny he had raised from seed of 'a Cowslip' contained both primroses and oxlips. Raven comments (p. 174): 'These plants certainly hybridise so easily that many botanists have treated them as a single species. But Ray must surely have been misinformed. Cowslip crossed with Primrose might produce the hybrid which superficially resembles the true Oxlip, but could not produce pure Primrose. Possibly Bobart had sown seed of the hybrid Oxlip and obtained the natural result, Primroses, Cowslips and hybrids.' Ray includes the example in his paper *On the Specific Differences of Plants* which he gave to the Royal Society in 1674. It is fascinating to find, three centuries later, that Oxford and Cambridge botanists still cooperate in studying the phenomena of hybridisation between the same three wild *Primula* species, and regularly use them as teaching examples when discussing the nature of specific distinction!

Ray's forced departure from Cambridge in 1662 brought to an end his narrowly academic career and deprived the University of the chance of rivalling Oxford in the scientific study of plants in England. It is clear from a letter to his friend the zoologist Willughby that Ray had in mind, not only a work on the whole British flora which he eventually succeeded in writing (the *Catalogus Plantarum Angliae* . . . published in 1670), but also, as he puts it, 'another catalogue which I will call *Horti Angliae*'. This list was to include all cultivated plants such as those already listed for the Oxford garden and the garden of Tradescant in Lambeth. We know that Ray had in fact a small 'botanic garden' by his rooms in Trinity, and there are references in his great three-volume work *Historia Plantarum* (1688–1704) to his having grown particular species there. Amongst these are relatively new and exciting plants from America such as the Tobacco plant '*Tabacco latifolium*' (*Nicotiana tabacum*) of which Ray says: 'it sometimes endures a winter as I proved at Cambridge'. There is indeed evidence (Raven 1950, p. 109) that he actually prepared such a catalogue of plants grown in Cambridge gardens and that this existed in manuscript form until at least 1739; it is unfortunately quite unknown now, and we are limited in our knowledge of Ray's Cambridge garden flora to the *Historia Plantarum* references which cover about forty species. As we see in chapter 4, it was more than a century later that the first catalogue of plants in cultivation in the Cambridge Botanic Garden was published.

How serious a deprivation was the absence of any facilities to study cultivated plants during his later career is well expressed by Ray in the preface to the *Methodus Plantarum*, published in 1682 (translation from Raven 1950, p. 193): 'I have not myself seen or described all the species of plants now known. I live in the country far from London and Oxford and have no Botanic Gardens near enough to visit. I have neither time nor means for discovering, procuring and cultivating plants.'

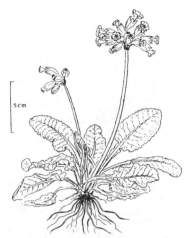

Primula veris L. Cowslip, Paigle   Deep yellow

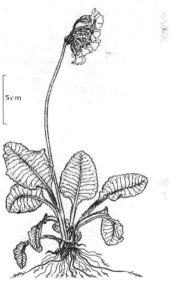

Primula elatior (L.) Hill   Oxlip, Paigle   Pale yellow

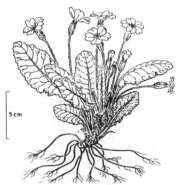

Primula vulgaris Huds. Primrose   Pale yellow

Figure 5   The three species of *Primula* (studied by generations of Cambridge and Oxford botanists from Ray to the present day).

To restrict the treatment of Ray to some account and assessment of his work on the local and the British floras only would be very misleading, but to expand this chapter to include, for example, a proper assessment of his achievements in the classification of plants would take us too far from our parochial theme. Perhaps we can be content to note that, in his essay on seeds in the *Methodus* (1682), Ray first makes that most important distinction in the natural classification of plants, namely between Monocotyledons and Dicotyledons. As Raven (1950, p. 195) puts it: 'The essay is literally an epoch-making piece of work: for its details he had the support of the greatest physiologist of the time: but the credit for discerning the significance of these details is Ray's alone.' In the *Methodus*, again to quote Raven: 'it could easily be argued that Ray in fact laid down lines of classification more in accord with genuinely scientific and evolutionary principles than those of his illustrious successor (Linnaeus)'.

Most of the years 1663–5 Ray spent travelling on the Continent, where he not only enjoyed his field botany, but also observed and recorded birds and fish, geology, peoples and their languages and customs. He was particularly interested in the state of learning in continental universities, and in fact studied anatomy during the winter of 1663–4 at the famous University of Padua. The contacts he made during these years gave him an exceptionally complete knowledge of the available published literature in Botany, and also ensured that he remained in touch with European science as it developed during the most productive years of his life. In all this, Ray's undoubted ability in languages must have been of outstanding importance. Not only was he skilled in Latin, Greek and Hebrew, but he also had sufficient proficiency in French and Italian, and a great interest in dialect words in English. Raven (1950) expresses the importance of Ray's publications in Latin as follows: 'He lived . . . in a time of transition. When he went to Cambridge, the old tradition which regarded Latin as essential for all serious and academic writing . . . was still unchallenged. "He is so ignorant that he cannot even write Latin without solecisms" is Ray's severest criticism of an opponent. His own mastery of the language gave a world-wide currency to his work. Gerard and Parkinson, writing in English, were unknown outside Britain: Ray was read by Tournefort and Hermann and Malpighi as readily as by his own countrymen or in the next generation by Linnaeus.' Conversely, it is fairly clear that publication in Latin was increasingly difficult and unpopular in England as the seventeenth century drew to a close, as Raven explains (p. 31): 'the universities . . . failed to develop any scientific interests except of the most desultory kind: the educated came to regard Latin as a dead language, of interest only to the

Figure 6 Frontispiece from Ray's *Methodus*, published in 1682.

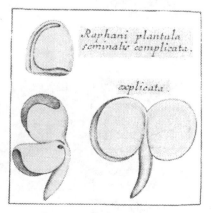

Figure 7 Drawings from the essay on seeds in the *Methodus*.

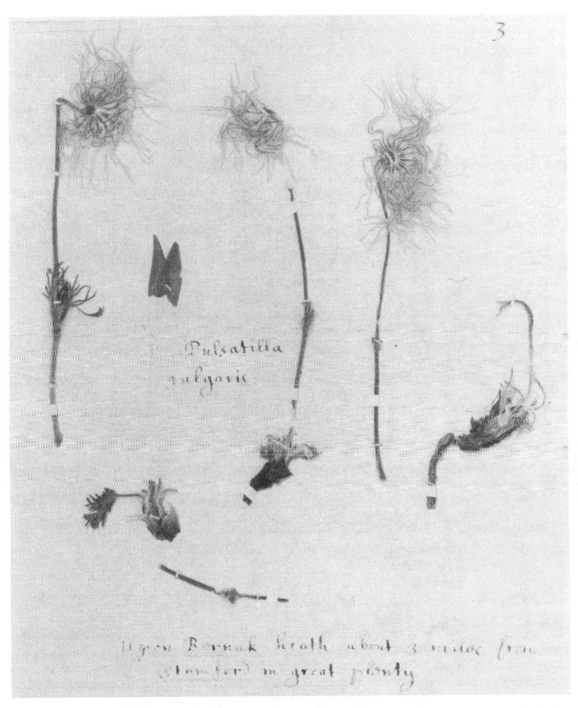

Figure 8   Specimen of the Pasque Flower, *Pulsatilla vulgaris* Miller, from Ray's own Herbarium, preserved at the British Museum (Natural History), London. The plant still grows at Barnack. It is interesting that the Latin name used a century before Linnaeus is the name in use in many Floras at the present day.

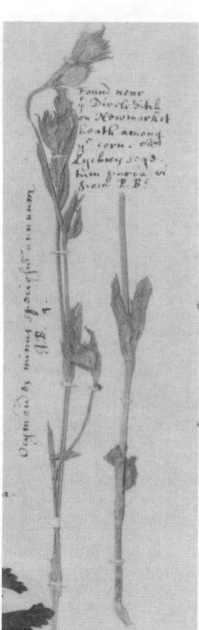

Figure 10 Specimen of the cornfield weed *Silene gallica*, found near the Devil's Dyke, the only Cambridgeshire specimen known in Ray's own Herbarium.

178 *CATALOGUS*

Park. ſylv. foliis diſſectis *Ger. emac.* ſylv. ſive Endivia multis dicta, folio laciniato, dorſo ſpinoſo *J. B. Cut-leaved wild Lettuce.* Ad ſepes & agrorum margines.

C. Lactūca ſylveſtris 2 *Ger. emac.* Lactucæ ſylveſtris ſive Endiviæ multis dictæ folio laciniato, dorſo ſpinoſo varietas *J. B. Endive-leaved wild Lettuce.* Cum priore, ſed rarior: Hæc non alia in re ab illa differt quàm foliis, quæ ei ne in caule quidem laciniata ſunt. Odore eſt minùs vehementi quàm prima ſpecies. Quæ de hac planta in C. C. tradidimus jam retractamus, & in Johnſoni ſententiam concedimus.

N. 1. Quæ de viribus Lactucæ ſylveſtris Veteres prodiderunt, huic plantæ (me judice) minùs conveniunt, Narcoticam eam eſſe & ſoporiferam, Opii vehemens & viroſus odor abundè convincit.

2. Quòd Lactuca ſylv. & omnia in genere opiata irritant potiùs quàm extinguunt appetitum ad Venerem, me monuit *D. Liſter*, ſaltem moderatè ſumpta, non ſecùs ac vinum ac alii generoſi liquores: quibus & viribus quodammodo reſpondere videntur.

3. Hujus herbæ incenſæ fumo animi causâ naribus excepto, ſe velut conſternatum & vertigine ſubita correptum, cóque plane modo affectum ſenſit *D. Liſter* quo ſolent illi qui Nicotianæ fumum haurire primò aggrediuntur.

C. Ladānum ſegetum quorundam ... Sideritis arvenſis anguſtifolia rubra *C. B.* arvenſis rubra *Park.* Sideritis *Ger. emac. Narrow-leaved All-heal, or Iron-wort.* Inter ſegetes. Floret à media æſtate in autumnum. Facultatibus cum Sideritide convenit, cujus etiam ſpecies cenſetur.

C. Lagōpus vulgaris *Park.* trifolius quorundam *J. B.* Lagopodium ſive Pes leporis *Ger.* Trifolium ſpicatum VI, i. e. arvenſe humile ſpicatum, ſive Lagopus *C. B. Harts-foot, or Harts-foot-Trefoil.* Inter ſegetes, maximè ſolo arenoſo.

The Wilisell hath since found it about Pancras church near London.

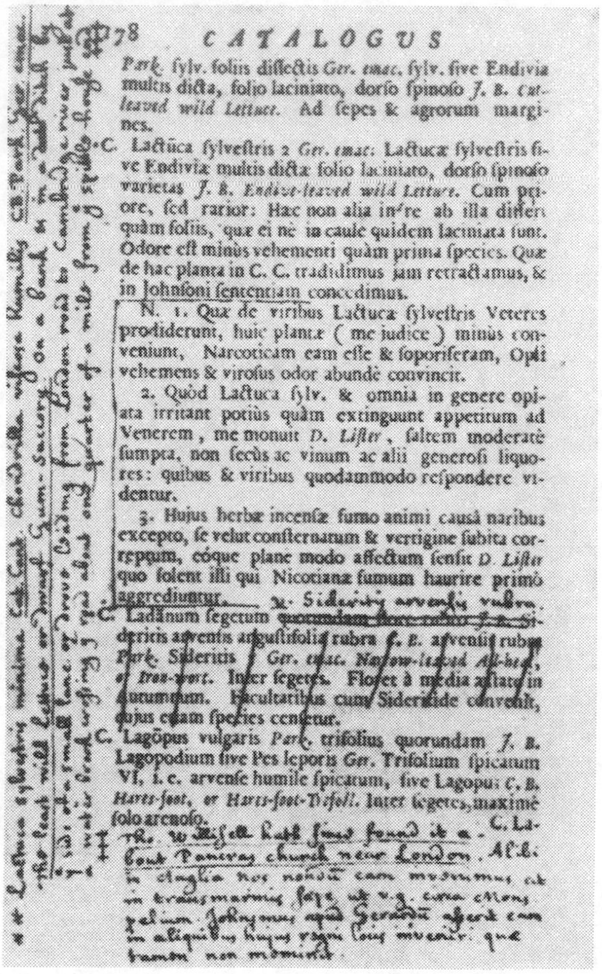

Figure 9 A page of the *Catalogus Plantarum Angliae* annotated by the author.

schools'. This is the scene we find when the next chapter of Cambridge botany opens.

A final word on Ray must be devoted to his most popular work, *The Wisdom of God Manifested in the Works of the Creation*, published late in his life in 1691, and quickly running to new editions. Unlike the purely biological books, this one is written in English. It remained an influential book throughout the eighteenth century; as Raven says, 'it supplied the background for the thought of Gilbert White . . .; it was imitated, and extensively plagiarised, by Paley in his famous *Natural Theology*, and more than any other single book it initiated the true adventure of modern science'.

3

Bradley and the horticultural tradition

Figure 11   Coffee plant in a glass-house of the type designed by Richard Bradley.

The eighteenth century was, in general, a time of stagnation in both the ancient universities. In Cambridge the promise of a rapid development of branches of natural science following the brilliant career of Newton at Trinity did not materialise, though the attitude to experimental science was generally less hostile than in Oxford, and individual contributions of real importance were made by Cambridge men throughout the century. In botany an outstanding contribution was that of Stephen Hales, who entered Corpus Christi College in 1696, took his M.A. degree in 1703 and was elected a Fellow of the College in the same year. An exceptionally enthusiastic and brilliant experimental scientist, Hales published in 1727 his *Vegetable Staticks*, a pioneer work devoted to the nutrition

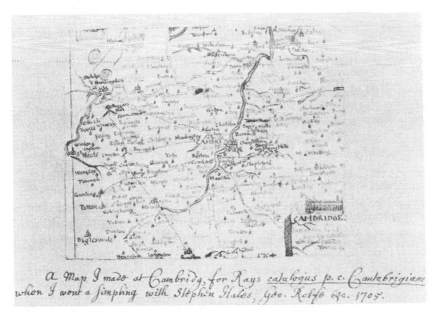

Figure 12   Map used by Stephen Hales and his friends when botanising in the vicinity of Cambridge using Ray's *Catalogus*.

of plants and the movement of the sap inside them and containing accounts of very many original experiments. Like Ray, Hales' career was that of an ordained clergyman: there was practically no place for such men to remain permanently as teachers and researchers in the University, and the normal career open to them was in the Church. A full biography of Hales was written by Clark-Kennedy in 1929.

How far the development of plant physiology after Hales was delayed by the unwillingness of the ancient universities to accept anything other than medically-subordinate 'herbal' botany is a fascinating question to which no clear answer can be given. The history of experimental plant physiology is certainly marked after Hales by a period of stagnation lasting for half a century, and this coincides with the relatively poor standing of experimental science in Oxford and Cambridge. Yet in both universities there were chairs of Botany before Hales' book was published . . . though in the case of Cambridge only just before.

At the time when Hales was conducting his experiments in Corpus Christi, there was a serious attempt to found a Garden in Cambridge. In October 1695 the then Vice-Chancellor, Dr Eachard, paid an official visit to London in connection with a projected Garden, and in the following year the King's Gardener, George London, visited Cambridge to 'measure the intended physick garden'. The site of this proposed Garden is not recorded, but a plan exists and has recently been acquired by the University Library.

Richard Bradley, first Professor of Botany in Cambridge, is a shadowy, enigmatic figure about whom we know surprisingly few facts – for example, even his date of birth is by no means certain, and therefore his age when he accepted the newly-created chair in 1724 remains doubtful. Until the present century, indeed, the opinion of his successors in the Chair, John Martyn and his son Thomas, had been generally accepted, namely that Bradley was a disreputable man, best forgotten, who had obtained the Chair by false pretences and disgraced the fair name of the University. Fortunately, though much of Bradley's life remains obscure, we now have a much more balanced picture of the man, largely owing to the researches of the Cambridge botanist H. Hamshaw Thomas (1937, 1952).

In his later paper, Thomas points out that Bradley was a most prolific writer, responsible for no fewer than twenty-four books published between 1714 and 1732, some of them running through several editions. Most of these works deal with agricultural or horticultural subjects, and many had considerable influence in both Britain and continental Europe, especially in France. Indeed the

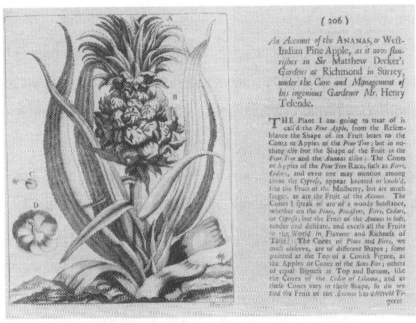

Figure 13   'An account of the Ananas or West Indian Pine Apple' from Bradley's *General Treatise of Husbandry and Gardening*.

claim has been made (Roberts 1939) that Richard Bradley was also the first 'horticultural journalist', producing a monthly *General Treatise of Husbandry and Gardening* in 1721–4.

The available facts about Bradley and his botanical publications have been conveniently brought together in Blanche Henrey's monumental work on *British Botanical and Horticultural Literature before 1800* (1975: vol. 2, pp. 424–54). The definitive biography of Richard Bradley, however, which Thomas hoped to write, was unfortunately never written, and Bradley continues to offer a fascinating challenge for the historian of science in need of an elusive but eminently rewarding subject.

Let us look at the facts which we have about the first Cambridge Professor of Botany. He became a Fellow of the Royal Society on 1 December 1712, being proposed for election by Robert Balle. Such meagre evidence as there is points to his being about 26 years old at the time of his election. As Thomas wrote (1952): 'at that time he must have occupied a decent position in society, he must have been considered of good intellectual standing, and able to pay his admission money and his annual subscription'. He had already announced the publication of his *Treatise on Succulent Plants* in an advertisement in 1710, although in fact it was to take him six years

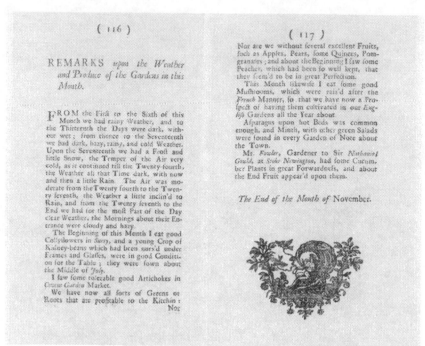

Figure 14   Monthly notes for November 1724, from Bradley's *General Treatise*.

to get the first of the five volumes of the *Historia Plantarum Succulentarum* published (1716–1727). Judged purely by this work alone, Bradley probably made a more significant contribution to botanical and horticultural science than either of his successors who, father and son, occupied the Cambridge Chair of Botany for over ninety years. Rowley (1954, 1964) who unravelled the complex bibliographic history of Bradley's books on succulents, sums up the work in the following words: 'Bradley laid the foundation of the study of succulents in England, and his influence was felt abroad also. His idea of a succulent [a plant 'not capable of a *hortus siccus*'!] tallies exactly with present-day likes and dislikes, whereas 100 years after him 'succulent' collections included saxifrages, cycads and mesophytes of unrelated habit and cultural requirements. His views on cultivation, too, were mostly nearer the truth than those of his followers, who tried to raise the plants in great heat and drought under glass.' Of the illustrations to the fifty succulents which Bradley gives, Rowley says: 'the majority are at once recognisable from the plates alone, and there can be no doubt that they were drawn by someone acquainted with living plants'. All but one of Bradley's succulents are still known in cultivation today.

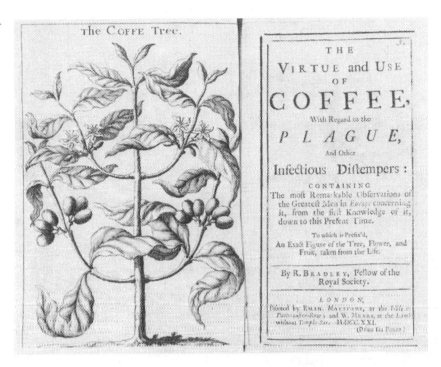

The Coffe Tree.

THE
VIRTUE and USE
OF
COFFEE,
With Regard to the
PLAGUE,
And Other
Infectious Distempers:
CONTAINING
The most Remarkable Observations of
the Greatest Men in Europe concerning
it, from the first Knowledge of it,
down to this Present Time.

To which is Prefix'd,
An Exact Figure of the Tree, Flower, and
Fruit, taken from the Life.

By R. BRADLEY, Fellow of the
Royal Society.

LONDON,
Printed by EMAN. MATTHEWS, at the Bible in
Pater-noster-Row; and W. MEARS, at the Lamb
without Temple-Bar. M.DCC.XXI.
[Price Six Pence.]

Figure 15   Title-page and illustration from Bradley's *The Virtue and Use of Coffee*,
1721.

In the year 1714 Bradley travelled extensively in Holland and
France, and made very effective use of his visits to the Physic
Gardens of Leiden and Amsterdam. One product of these visits was
his *Short historical account of coffee*, an illustrated treatise based upon
his study of *Coffea arabica* in cultivation in Amsterdam, which was
presented by Balle to the Royal Society 'from Mr Bradley' in April
1715. Although there is no statement to clarify in what capacity
Bradley worked with Balle, there is much indirect evidence (for
example, in letters from Bradley to James Petiver preserved in the
Hans Sloane manuscript at the British Museum: see Tjaden 1973–6)
that Bradley was able to use and enjoy Balle's garden at the great
mansion, now long destroyed, of Campden House, Kensington.
Robert Balle was a merchant at Leghorn who was the owner or,
more probably, the tenant of Campden House for several years up
to 1719; described by a contemporary as 'that ingenious encourager
of vegetable nature', he seems to have been a rich amateur gardener
much interested in experiment.

In such an atmosphere Bradley obviously throve, and published
the first of his many works on horticulture, the *New Improvements of
Planting and Gardening* (1717–18). Already, however, there is

Figure 16   Illustration from Bradley's *Philosophical Account of the Works of Nature*, 1721, showing Algae, Fungi, and Cacti: part of an 'evolutionary series' of plants.

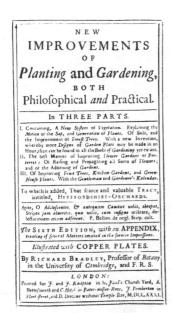

Figure 17   Title page and frontispiece from Bradley's *New Improvements of Planting and Gardening*, sixth edition, 1731. Like many of Bradley's books, this ran to several editions, some of them posthumous.

evidence of money troubles, and indeed it was only the intervention of James Brydges which saved him from a debtors' prison in 1717. By 1720 Bradley's fortunes seem to have suffered a reversal from which he never recovered. In a letter to Hans Sloane dated 23 June 1722, Bradley alludes to 'the unfortunate affair at Kensington whereby I lost all my substance, my expectations and my friends', and it is clear that his 'substance' included not only money and position, but also his living plant collection, much of it originally acquired from the Amsterdam physic garden. The letter ends, rather pathetically:

I have some friends at Court who do not care I should go abroad tho my inclinations are for it, even into the most dangerous country; but to live upon expectations at home is as bad as it can be to venture ones life among the savages abroad; but to free my self from both these I would chuse to have a garden of experiments for general use, such as I should have accomplisht if I had not had the Kensington misfortune & by that means I might gain an improving settlement and I hope do my country some service without restraint of booksellers. This Sr. I submit to your consideration & humbly beg your advice.

By October 1723 we find Bradley asking Sloane to promote his candidature for the Chair in Oxford 'for I have made some interest there for the Professorship of Botany which will now be put on a

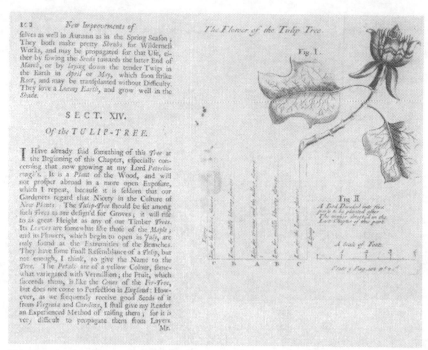

Figure 18   Description and illustration of the Tulip-tree, *Liriodendron tulipifera*, from Bradley's *New Improvements*.

new foot and the physick garden brought into order'. He was unsuccessful, and Oxford appointed Gilbert Trowe, who left no mark at all on the subject of botany. The following year, however, the University of Cambridge elected Bradley Professor of Botany by a Grace passed by the Senate on 10 November 1724, a post which he held until his death in 1732. The grace alluded to Bradley's promise to expend 'both funds and energy' on establishing a Botanic Garden. It specified neither duties nor emoluments for the Professor, but in this respect was by no means unique (Winstanley 1935 gives a wealth of detail about the Chairs in the University and their holders in the eighteenth century, and makes very good background reading). Whether Bradley himself thought that the election to the new Chair of Botany in Cambridge would end his troubles we have no means of telling. His detractors later accused him of obtaining the appointment 'by means of a pretended verbal communication from Dr Sherard to Dr Bentley, and pompous assurances that he would procure the University a public botanic garden by his own private purse and personal interest' (T. Martyn (ed) 1770, p. xliv). That the notorious Dr Bentley, Master of Trinity College, was involved in getting Bradley the Chair seems quite probable; for along with many other Cambridge men, he appears as a subscriber to Bradley's *Philosophical Account of the Works of Nature* published in 1721. Indeed Bentley's interest in Bradley may well have been respectably intellectual, for he had earlier persuaded the Fellows of Trinity to make for the new Professor of Chemistry, John Francis Vigani, 'an elegant Chemical Laboratory' in the College, and was obviously interested in furthering practical science as well as promoting, as he put it, 'the dignity of the College'. Monk, the nineteenth-century biographer of Bentley, describes (2nd edn 1833, 1 : 201–205) how Bentley 'had the satisfaction of founding in Trinity College a school of natural philosophy of singular eminence. . . . It was at this time (1708) his favourite object to make Trinity College the focus of all the science in the University'.

To pretend, as did the Martyns, that using influence and 'knowing the right people' was somehow unusually disreputable and dishonest of Bradley does not bear much examination. It was probably true that most Professors owed their chairs to such practices. Even less does Bradley's 'hollow claim' to procure for the university a public botanic garden seem to be a legitimate ground for complaint, for there is ample evidence that Bradley wanted more than anything else a 'physic garden' where he could pursue his botanical and horticultural studies. Perhaps we should now look more closely at what Bradley had in mind, and see why he failed.

In the Library of the Cambridge Botany School is preserved a manuscript volume of twelve lectures prepared by Bradley in the early months of 1725 and obviously intended to be delivered by him soon after his appointment. We do not know whether he ever delivered any of these lectures, but he is certainly maligned by Gorham, in his memoirs of the Martyns (1830) who blamed Bradley for 'neglecting' to deliver lectures. As Thomas (1937) explains, it is the *content* of the manuscript lectures which gives 'a possible reason for Gorham's charges'. Bradley's lectures are described on the title-page as 'explaining the principles of vegetation' on which the 'arts of husbandry and gardening' ought to be based. The first lecture, logically enough, deals with the different kinds of soils 'so that we may know the use of earth in vegetation as the foundation of all our work'. Bradley explains: 'However useful and necessary the arts of Husbandry and Gardening have been accounted in all ages, we have hardly found any of the writers on those subjects have given themselves the trouble of laying before us the principles or rudiments on which those great and beneficial works ought to be founded.'

What interested Bradley was clearly quite different from the botany of Turner and Ray: as Thomas puts it, 'Bradley's botany, as may also be seen from the many [other] works which he published,

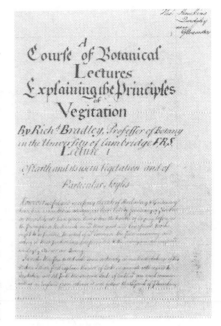

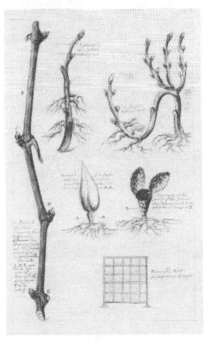

Figure 19 Title-page and illustrations from the manuscript course of lectures by Bradley preserved in the Botany School.

was the botany of Malpighi, Grew and Hales, and not the botany of Ray.' It is in fact applied and experimental, not systematic and nomenclatural, botany. It seems that Bradley had no interest in, or knowledge of, the British or European floras, and it is a striking fact that no single Latin name of any plant appears in his twelve manuscript lectures! No wonder that this man, who apparently had no university education, was unacceptable to the traditional, classically-dominated ancient University. It was clearly no part of a University education to learn how to graft your vines or prune your fruit-trees: such traditional crafts were performed by the gardeners, paid for by the 'young gentlemen' or their fathers who sent them to Cambridge to learn 'a real profession' such as the Church, Law or Medicine.

In fact, Bradley *did* succeed in adapting himself to the requirements of the University system, and delivered in 1729 a course of lectures on *Materia Medica* which he published in 1730. (It is interesting that he used for these lectures the Vigani collection of *Materia Medica* still preserved in the Library of Queens' College, and his signature appears on two of the papers which line the specimen trays.) These published lectures were satirically criticised by John Martyn (writing under the pseudonym of 'Bavius') in the *Grub Street Journal* (1730): 'It was particularly obliging of our worthy Professor to print these lectures: seeing no more than three or four of our students had the pleasure of hearing them read.' He continues: 'The book before me gives me a great idea of his genius. For he scorns, like the vulgar writers on the *Materia Medica*, to copy in a servile manner from other authors; but, bravely deviating from the beaten road, gives us, almost in every page, something equally new and surprising.'

In this last sentence, Martyn's sarcasm rebounds on himself. Seen from our position, the young upstart Martyn, fashionable and conservative in his views, seems to condemn himself, and Bradley's stature is only improved. Is it necessarily a condemnation of Bradley that he puts his own ideas and experience into an otherwise very pedestrian *Materia Medica* course?

Perhaps the social gulf between Richard Bradley and the Cambridge dons was an even greater barrier than the content of his expert knowledge and the gaps in his classical education. We can sense what trouble this must have caused when we read the letters Bradley wrote to his friend the apothecary-botanist Petiver, several of which were written from Holland in 1714 (Tjaden 1973–6). The racy style, the frankness with which he reports his money troubles and his posing as a medical doctor, all reveal Bradley as a rather engaging but unreliable 'man of the world'. Thus, writing from The Hague in July 1714, Bradley says: 'If I should be gone before you answer this which I think not to do because of the Rhino [money from England!] it will follow me and come safe. In a word if Money does not come I must turn another Ovid and Metha-morphose Coats and Waistcoats into Guldens or else with Icarus's wings fly "til I am melted body and boots".' Letters written by his successors John and Thomas Martyn seem stiff and formal by comparison, belonging to another world.

When we turn to Bradley's efforts to secure a Botanic Garden for Cambridge, we find the facts are very difficult to ascertain. There is no doubt from his published works what Bradley wanted, but it is by no means clear how he went about trying to get it, and indeed why he failed. Let us look at what he wrote about the Garden he wanted to establish in Cambridge. He writes twice on this subject: the first reference (1725) occurs in the Preface to his book entitled *A Survey of the Ancient Husbandry and Gardening*, and the second (1730) in the Introduction to his published *Materia Medica* lectures. The earlier passage is so explicit – and so little known – that it is worth quoting *in extenso*. After explaining that in ancient Greece and Rome 'husbandry [i.e. agriculture] was accounted a study so extremely beneficial to the commonwealth, that persons of the highest rank and figure did not only promote the practice and improvement of it among the common people, but took a pride to distinguish themselves by such new inventions and contrivances, as might add anything to an art of so general advantage,' he continues as follows: 'I think Britain might yet be brought to a much greater perfection in agriculture than it is at present, if our farmers had opportunities and judgment to try experiments, or had some fixed place, where they might see examples of all kinds of husbandry, as a School for their information.' This, he says, 'I hope to compass, as soon as a Physic Garden is completed at Cambridge, where, besides collecting such plants as are used in physic, and choice vegetables from foreign countries, a little room may be spared for experiments tending to the improvement of land, which may be the means of increasing the estate of every man in England; for in such an undertaking every kind of soil must be used, and every situation imitated.'

After this strong defence of what we would nowadays call the 'Research Area' of the Botanic Garden, Bradley sets out a reasoned case for the planned Garden growing and displaying for teaching the plants used in medicine, so that 'the young gentlemen who study physic at the university will then have opportunity of know-ing the plants, and even the drugs they are to use'. He follows this with a remarkable plea, based upon what he had seen in Amster-dam, for the Garden as a place where exotic plants of economic importance can be received, studied, cultivated and propagated to start new agricultural enterprises in other parts of the world. He cites the history of the introduction of coffee 'which at first they cultivated at Batavia . . . and brought trees of it to Amsterdam; where, after a little time they raised several hundreds and sent them to Surinam and Curasau, in the West Indies, from whence, I am told, they receive a good freight of coffee every year'. He concludes this passage: 'When I . . . reflect upon the state of our American plantations, and our extensive trade, I can see no reason but that we may render them more advantageous than they are at present, by sending to them many plants of use, which will grow freely there, and may be collected and prepared for them, in such a Garden as I speak of.' A century and a half later this was precisely the role played by the Royal Botanic Gardens at Kew in relation to tropical and subtropical crops in the British Empire.

The whole case made by Bradley for the Physic Garden in Cambridge was in terms of a practical, applied science. The idea of 'pure' learning, science for its own sake, is conspicuously absent. This may, of course, have been a political decision; Bradley had to judge what would best commend his project to the 'noblemen and gentlemen' who had 'promised to contribute towards a Garden at Cambridge, and merit the honour of being enrolled in the list of its "Founders"', and we have plenty of evidence from other published writings that he had a very lively scientific curiosity which did not need any other justification.

The Introduction to the *Materia Medica* lectures, given in 1729, but published in 1730, does not repeat the three-sided justification of his desired new Garden, but simply reports progress in establishing it: 'I have directions from several of my friends, who are persons of quality, and honour, to find out a proper piece of ground in this University, to be purchased for a Physic Garden, and put in such order as may render it both useful and ornamental.' He says further: 'I have at length fixed upon one which, if the Gentlemen of the College it belongs to, will consent to part with on reasonable terms, will in all probability be purchased, and secured to the University by Act of Parliament.'

Exactly what piece of land Bradley had in mind is quite unknown, nor is it at all clear whether he really had any substantial backing from friends who were 'persons of quality and honour'. Thomas Martyn tells us that in 1731 his father 'entertained sanguine hopes that a Botanic Garden would have been founded at Cambridge by Mr Brownell of Willingham, a gentleman possessed of a handsome fortune and a great lover of the science. But though there passed many conferences upon the subject between him and the vice-chancellor, Dr Manson, then Master of Bennet College, and now Bishop of Ely, and Dr Savage, the Master of Emmanuel

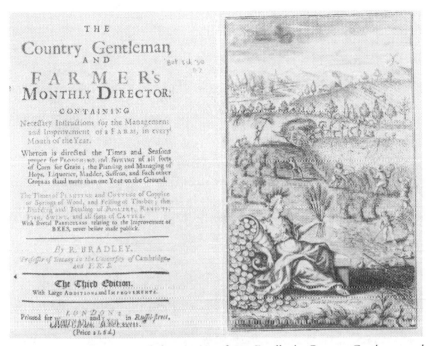

Figure 20   Title-page and frontispiece from Bradley's *Country Gentleman and Farmer's Monthly Director*, third edition, 1727.

College, though the ground was actually pitched upon, and Mr Miller was called in, to deliver his opinion concerning the design into execution: yet it was dropped, and Mr Brownell's estate was diverted into another channel.' It seems probable, though unproved, that 'a worthy gentleman of this County', whom Bradley mentioned in the Introduction as a potential benefactor of importance, was in fact Mr Brownell.

Whatever the exact cause of the last-minute withdrawal of Mr Brownell's support, it is reasonable to suppose that the final collapse of the scheme was a very severe blow to Bradley. According to Thomas Martyn (1770), by 1731 Bradley 'was grown so scandalous, that it was in agitation to turn him out'. He died on 5 November 1732. To the Martyns who followed him, Bradley was a chapter in history best forgotten. As we now see it, he looks more like a pioneer experimental and applied scientist who had the misfortune to be appointed to the Chair of Botany at the wrong time. Perhaps his character accentuated the opposition to this 'uncultured scientist', but we cannot judge, for we have only Thomas Martyn's word for it that Bradley was 'scandalous'; it may be significant that, although John Martyn publicly attacked him, Bradley never replied in print.

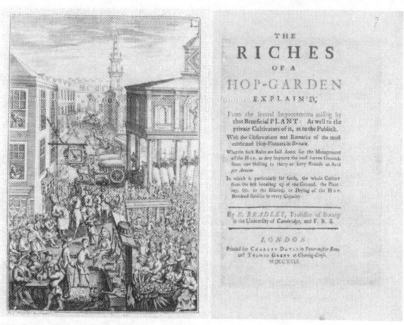

Figure 21   Title-page and frontispiece from Bradley's *The Riches of a Hop-Garden Explained*, 1729.

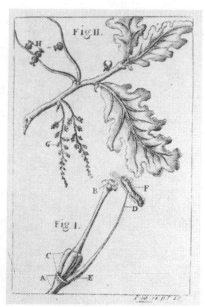

Figure 22   Illustration of pollination from Bradley's essay on pollination in *New Improvements of Planting and Gardening*, 1717.

Before we turn to the careers of the Martyns, a brief review of the scope and range of Bradley's published works will perhaps indicate something of his exceptional worth and reputation. As we have already seen, most of them are devoted to practical horticulture and agriculture. In these works we find a wealth of practical experience and a keen interest in experiment covering a surprisingly wide range of biological subjects. He is revealed as a pioneer in the study of plant diseases and human epidemics, in pollination, species-hybridisation and plant breeding, in variegation and its relation to virus infection, and even in studies of ecological productivity.

In his views on what we would now call plant breeding, as Rowley (1964) has pointed out, Bradley may well reveal his most original thinking. Not only did he report what is generally recognised as the first clear case of an artificially-made garden-plant hybrid – the cross between sweet william and carnation (*Dianthus* species) made in his friend Thomas Fairchild's nursery, and described in 1717 – but he constantly advocated controlled hybridisation as a horticultural practice. As with the germ theory of epidemic disease (which Bradley advocated, and the medical profession did not accept for another century – see Williamson, 1955), so scientific plant breeding had to wait for the nineteenth century to develop. A

convenient place to find Bradley's views expressed is under the
entry 'Vegetation' in his *Dictionarium Botanicum* (1728).

As Thomas (1952) says:

Although Bradley's many books show a large amount of repetition
they are full of interesting information, and they display a remarkably
modern outlook. He fully realised the importance of soil and climatic
factors in determining the growth of plants, and we constantly find
him advising his readers to try experiments. I cannot help feeling that
Fate has treated very badly a man of considerable originality, ability
and industry. The significance of his work has never been properly
appreciated, but when his works are read and studied from the
historical standpoint, he will rank as one of the outstanding British
plant-biologists.

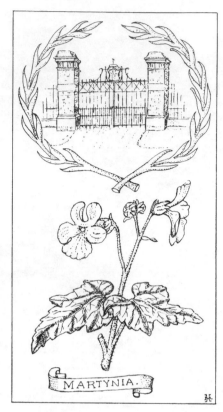

Figure 23 *Martynia* and the gates of the Chelsea Physic Garden.

4

The Martyns and the Linnaean tradition

John Martyn, Professor of Botany 1733–1762

With the death of Bradley in 1732, the way was clear for his rival, John Martyn, who had already moved into the Cambridge botanical scene and established himself as a natural successor. Martyn was born in 1699, son of a wealthy merchant trading in London and Hamburg; his first botanical interests seem to have developed out of herborising with young friends on excursions organised by the Society of Apothecaries and centred on the Chelsea Physic Garden. By 1721, his group had grown so interested that the 'Botanical Society' was formed, with John Martyn as its first Secretary, and the famous German botanist Dillenius, newly-installed in Britain, as its first President. A fascinating account of this Society, which may be the first formally-constituted Botanical Society in Britain – or even in the world – is given by Allen (1967). It is clear that the Society was young, enthusiastic, and largely composed of 'medico-botanists': one-third of the identifiable members were apothecaries, one-third physicians and most of the others surgeons or surgeon-apothecaries. As Allen says: 'the members and their immediate families form a fine cross-section of the intellectual middle class at this period, the teaching profession alone being under-represented'.

Martyn at the time was working in his father's office in Cheap-side as a clerk. He seems quickly to have found a talent for teaching botany, and developed a course of lectures which he originally gave to the Society, and soon to a wider audience in London. Out of these lectures grew the invitation to lecture in Cambridge, perhaps arranged by his friend Thomas Richmond, who entered St John's College in 1722, and seems to have subsequently studied medicine in London. We learn that 'more than twenty scholars' invited Martyn, and he duly gave his course of lectures in the Anatomy School in Cambridge in 1727. The introductory lecture of this course was published in 1729: its style and content form an interesting contrast to the *Materia Medica* lectures of Bradley, and were clearly designed to teach young medical students what we would now call plant morphology as a preliminary to their learning to identify individual species. The booklet is dedicated to 'Dr Christopher Green, Regius Professor of Physick in the University of Cambridge', to whom John Martyn explains, and almost apologises, why the book is in English:

> Some perhaps, tho' I think with little reason, will find fault with me for writing in the vulgar tongue. Had these lectures been delivered from a publick Chair; then custom indeed would have obliged me to express my self in the learned language. But, as they were designed only for the private instruction of some gentlemen, who were pleased to commit the direction of their studies in botany to my care, I thought that most proper, which was the native tongue of us all. It would have been rather easier to my self, to have spoken in Latin; most of the books of botany being written in that language . . .

We should not dismiss this last sentence as pretentious boasting. It is more than likely true: in presenting his plant morphology in English, Martyn was attempting to find the vernacular equivalent for a set of classical technical terms which were already largely standardised in the writings of European botanists. These terms (e.g. *flos, calyx, radix*) are either Latin, or Greek with a Latin form, and Martyn gives copious notes in explanation, quoting as appropriate from previous botanical authors in Greek, Latin, French and English. It is interesting that Nehemiah Grew's great work the *Anatomy of Plants*, published in 1682, is the only authority whom he quotes in English; there are several passages from the works of Ray, but these, of course, are in Latin.

A feature of the booklet which must have greatly increased its value was the copious illustration. Martyn explains that, in his view, the absence of illustrations from the works of Ray, and Caesalpinus before him, was responsible for their relative neglect in favour of the works of Tournefort. The labelled drawings of flowers in Martyn's booklet must have been a great help to his students, and it comes as a surprise to see how much comparative floral morphology is here presented at a time when the function of flowers was still unclear. Martyn's definition of a flower, arrived at after discussion over two pages of text, comes out rather well and, indeed, cannot be much improved on today: 'a flower, *flos*, is the organs of generation of both sexes adhering to a common placenta, together with their common coverings; or of either sex separately with its proper coverings; if it have any'.

John Martyn's election to the Royal Society took place on 30 March 1727. He had in the previous year published his first work, a twenty-page booklet on medicinal plants entitled *Tabulae Synopticae Plantarum Officinarum* and dedicated to Sir Hans Sloane, and

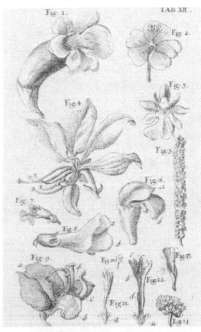

Figure 24 Illustrations of floral structure from John Martyn's *First Lecture of a Course of Botany*, 1729.

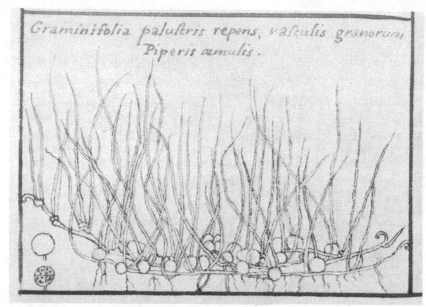

Graminifolia paluftris repens, vafculis granorum, Piperis amulis.

Figure 25   Unpublished illustration of *Pilularia* by John Martyn, preserved in a bound manuscript Flora in the library of the Botany School. Note the cumbersome pre-Linnaean name.

followed this in 1727 with *Methodus plantarum circa Cantabrigiam nascentium*, for the use of his pupils in Cambridge. In the Botany School Library is preserved an interleaved copy of Ray's *Catalogus*, copiously annotated by John Martyn, and bound together with a copy of the *Methodus* also interleaved and annotated by the author – a unique 'working document' which has survived miraculously to the present day.

John Martyn's most important publication, the *Historia Plantarum Rariorum*, appeared in five 'decades' (parts containing ten plants) between 1728 and 1737. Henrey says 'it bears the *imprimatur* of the Royal Society, with the name of Sir Hans Sloane as President, and is dedicated to the President, Council and Fellows . . . The work is of particular interest as it contains some of the earliest examples of colour-printing from a single metal plate. These plates were executed by Kirkall in a mixture of line-engraving – for the titles and coats-of-arms of the dedicatees – etching, and mezzotinting.' This impressive work was extremely expensive to produce, and was discontinued after the fifth decade. The botanical descriptions are accompanied by horticultural notes supplied by Philip Miller, Curator of the Chelsea Physic Garden whose help Martyn graciously acknowledged in the Preface: 'De cultura autem, cuius in praxi me parum esse versatum ingenue fateor, amicum jucundissimum Millerum semper consulo.'

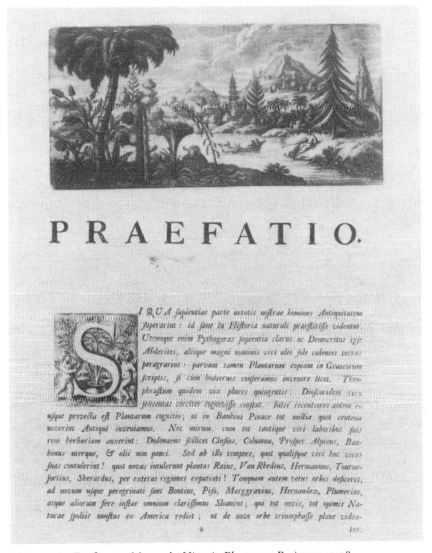

Figure 26   Preface to Martyn's *Historia Plantarum Rariorum*, 1728.

In May 1730 Martyn was admitted to Emmanuel College, where he kept five terms. It was his intention to proceed to a degree in medicine, but he abandoned this plan, and settled down to practise medicine, first in Bishopgate, and later, after his marriage in 1732, in Chelsea. Thomas Martyn records that at Chelsea his father had 'practised physic with tolerable success, and great reputation, for above twenty years; till increasing infirmities forced him into retirement'.

Figure 27   Illustration of a *Pelargonium*, newly introduced into cultivation from South Africa, in Martyn's *Historia*. Note Emmanuel College crest.

Figure 28   Fruit of *Martynia* (*Proboscidea*).

Martyn's election to the Chair of Botany took place on 8 February 1733. Apparently the affair did not go too smoothly; there were originally two rival candidates, who had the advantage of being resident members of the Senate, and even when support for them had declined and they had withdrawn, there was still a suspicion that 'our good friend, the Master of Christ's intended to stop [the] Grace in the Caput' . . . on the suspicion that Martyn was unwilling to 'take the oaths'. A letter from Martyn written in December 1732 to his friend Richard Arnold, a Fellow of Emmanuel, who was promoting his name for the Chair, is very revealing of Martyn's attitude. He says: 'I thought the University would never have made any difficulty about giving me *an empty title* in a science which I had restored, after it had been totally lost among them, and had continued to teach for six years with much labour and little profit . . . I find it objected of me, that I have left the University, which you can refute, for you know that my name is still on the boards at Emmanuel . . . How inconvenient soever it may be to me in the following of my practice here, which absence must necessarily injure, I shall nevertheless endeavour to serve the University *whenever it can be done without the greatest detriment to my private affairs.*' (My italics.)

It is difficult to escape the conclusion after reading this passage that John Martyn was already, even before his election, treating the Chair of Botany as a richly-deserved distinction for services already rendered, rather than a great career for the future. As we have seen, the exciting promise of a Botanic Garden had come to naught in the year before Bradley's death, and there is no evidence that Martyn made any further attempt in this direction after his election. Indeed, we are told by his son that 'in the year 1735 he read his last Course of Lectures in Botany at Cambridge; labouring under great disadvantages for want of a Botanic Garden, and not finding sufficient encouragement to warrant so long a neglect of his practice as the Course must necessarily occasion'.

John Martyn's contribution to the Cambridge botanical tradition took place, therefore, mainly before he became Professor, and in the area which Ray had pioneered. He obviously enjoyed instructing the young in field botany, and was concerned to make sure they understood the basic morphology on which identification was necessarily founded. He was an observer and recorder, and seems to have had no interest in experiment: the very opposite of Richard Bradley. Yet there was one thing which they both shared, namely a desire to found and develop a Botanic Garden. We have seen what Bradley wanted *his* Garden to do: to find out what John Martyn wanted we must wait thirty years, to see the Garden established just at the time when he relinquished the Chair, and was succeeded by his son Thomas. In the University Herbarium, at the Botany School, are preserved a number of specimens prepared by John Martyn in the period 1729–31; their provenance is not clear, but from the inscription 'H.C.' on them it is tempting to suppose that he grew them in 'Hortus Cantabrigiensis', an unrecorded predecessor of our present Garden, perhaps in Emmanuel College? Most of Martyn's Herbarium, which he left together with his botanical library to the University, consisted of foreign plants which he had obtained from friends, among them the botanist Houstoun, who collected in the West Indies and mainland America, and named after him the genus *Martynia*.

Before we leave John Martyn, some evaluation of his published work should be attempted. His name as an author is associated with translations of the works of Virgil, especially the *Georgics* and the *Bucolics*, where he was able to use his botanical expertise in attempting to identify the plants named in these works. His other important books are the *Historia Plantarum Rariorum* which we have already discussed, and a two-volume translation of Tournefort's *History of Plants* (1732). Remarkably few of his communications to the Royal Society are on botanical subjects, the most interesting being the latest (1754) on the dioecism of holly (*Ilex aquifolium*). From his published works Martyn is revealed as a moderately successful professional man with a good classical education and a safe position in eighteenth-century society, with leisure to pursue his literary researches and to correspond with a wide circle of acquaintances, including Linnaeus himself, who met Martyn during his short visit to England in 1736, and who presented him with copies of his *Flora Lapponica*, *Genera Plantarum* and *Critica Botanica*. (All these copies, autographed by Linnaeus, are in the Botany School Library today.)

From 1735 to his retirement from the Chair in 1762 Martyn seems to have shown no interest in Cambridge Botany, nor to have made any attempt to give lectures in the University. Although, as we have seen, absentee Professors were not rare in eighteenth-century Oxford and Cambridge, it is not easy to exonerate Martyn from a charge of indifference or even indolence.

Figure 29   Illustration of *Passiflora* from Martyn's *Historia*.

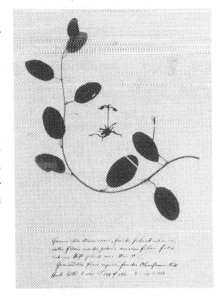

Figure 30   Specimen of same in Martyn's Herbarium preserved in the University Herbarium in the Botany School. Note 'H.C.' and date, September 18, 1731.

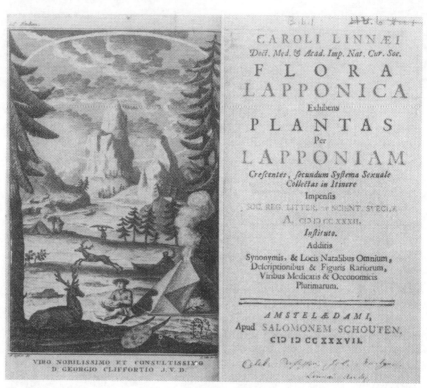

Figure 31   Title-page and frontispiece of John Martyn's copy of Linnaeus' *Flora Lapponica*, 1737, autographed and presented by the author.

Thomas Martyn, Professor of Botany 1762–1825

The smooth succession of John Martyn's son Thomas to the Chair of Botany on the retirement of his father in 1762 looks suspicious and, if we are to believe a revealing quotation attributed to Thomas in his later years, it must have involved collusion. 'I succeeded', he is reported as saying, 'because I canvassed with my father's resignation in my pocket before anybody was aware of it.' What more natural than that Thomas was groomed for the job, and slotted into it at the ideal age of 26 years?

Thomas Martyn was to hold the professorship for over sixty years. In his favour several things can be said, the most obvious being that he took his teaching duties seriously and conscientiously for the first half of his long academic career. Yet we cannot escape the feeling that he was a fortunate man to whom everything had been easy, but whose talents were mediocre. Let us look a little more closely at his education and his achievements, both as a teacher and a writer.

Born in 1735 in the house in Church Lane, Chelsea, where his father practised as a doctor, young Thomas was privately educated nearby, and his childhood until his mother's death when he was thirteen years old must have been singularly free from trouble. Gorham, biographer of the Martyns, says (1830, p. 86): 'he lived, during this period, with great domestic comfort; and the foundation of his future eminence was, undoubtedly, laid in a solid education, imparted by a competent tutor under a father's vigilant eye'. In his old age Thomas recalled his memories of visiting Sir Hans Sloane: 'The condescension of the venerable and amiable old gentleman to me, when a schoolboy, will never be forgotten by me. His figure is, even now, presented to my eye, in the most lively manner; as he was sitting fixed by age and infirmity in his arm-chair. I usually carried a present from my father of some book that he had published, and the old gentleman in return always presented me with a broad piece of gold, treated me with some chocolate, and sent me with his librarian to see some of his curiosities.'

When Thomas was seventeen years old, his father arranged for his admission to his old College, Emmanuel, where he worked diligently if not brilliantly and took his B A in 1756 Reading between the lines of Gorham's excessively flattering biography, we get the impression that Thomas was interested in the factual and descriptive, but had neither taste nor ability for mathematics. He now, at twenty years old, 'applied himself principally to preparation for Holy Orders', and we learn further from Gorham: 'that he might pursue his studies more at leisure, his father indulged him with residing much in College, till he was of age to take Deacon's Orders'. So Thomas Martyn became a 'College man'. The next step in his career came with the offer of a Fellowship at Sidney Sussex, conveyed to him in a letter from Rev. H. Hubbard, his Tutor at Emmanuel, on 16 January 1758. This letter, given in full by Gorham, contains a revealing passage which sets out what the Fellowship examination entailed: 'some knowledge of Hebrew is necessary: the rest of the examination is usually in Philosophy, Aristotle's Rhetoric, some part of the first six books of Homer, and Virgil's Georgics'. The examination seems to have presented no difficulty to young Thomas, whose private education had given him a 'thorough grounding in the classics', and he was elected a Fellow of Sidney in May 1758. In the following year he took his M.A. degree and was ordained priest by the Bishop of Lincoln. By 1760 he had become joint Tutor of Sidney with another Fellow, John Hey, with whom he worked, very amicably, for fourteen years until his marriage in December 1773 to Martha Elliston, the Master's sister. They resided at Thriplow till 1776, then in several other places until he finally 'retired' in the Rectory at Pertonhall, Bedfordshire in 1797.

22 *Plantae Cantabrigienses.*

Monoecia Polyandria.	METH. CANT.	CAT. CANT.
Sagittaria sagittifolia	Sagitta, p. 56. 1.	Sagitta, 1.
β	2.	2.
Poterium sanguisorba	Sanguisorba, p. 20. 1.	Pimpinella, 2.
Quercus Robur	Quercus, p. 115.	Quercus
Carpinus Betulus	Carpinus, p. 114.	Betulus
Corylus Avellana	Corylus, p. 114.	Corylus
Dioecia Diandria.		
Salix amygdalina	Salix, p. 116. 4.	Salix, 4.
fragilis	3.	3.
purpurea	10, 11.	9, 10.
Helix	5.	5.
repens	6.	1, App.
β	12.	8.
caprea	8.	7.
β	7.	6.
viminalis	9.	8.
alba	1, 2.	1, 2.
Tetrandria.		
Viscum album	Viscus, p. 117.	Viscum
Myrica Gale	Gale, p. 115.	Elaeagnus
(1) Aquifolium vulg.		
Pentandria.		
* Cannabis sativa	Cannabis, p. 11.	Cannabis
Humulus Lupulus	Lupulus, p. 10.	Lupulus
Hexandria.		
Tamus communis	Tamarum, p. 61.	Bryonia, 2.
Octandria.		
Populus alba	Populus, p. 115. 1.	Populus, 1.
nigra	3.	3.
tremula	2.	2.
Enneandria.		
Mercurialis perennis	Mercurialis, p. 11. 1.	Cynocrambe
annua	2.	Mercurialis, App.

Dioecia

(1) I have ventured to place the *Holly* here from the observation made by Dr *Martyn* in the Philosophical Transactions, Vol. XLVIII. p. 613. I have observed a great number of wild plants myself, and never found any but such flowers as he has described; that is, female with four abortive chives, and male flowers: I never could see a perfect hermaphrodite flower, nor male and female on the same plant. If the abortive chives should be thought sufficient to render what I call the female an hermaphrodite flower, the *Holly* will belong to the next class of *Polygamia Dioecia*; but if Dr *Watson* be right in his remark on Dr *Martyn's* observation, this plant ought to be placed in the class of *Polygamia Trioecia*.

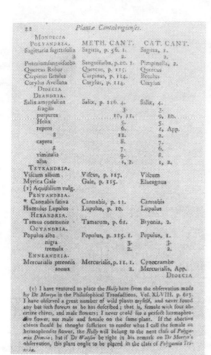

Figure 32 Page from Thomas Martyn's *Plantae Cantabrigienses*, 1763, set out according to the sexual system of Linnaeus.

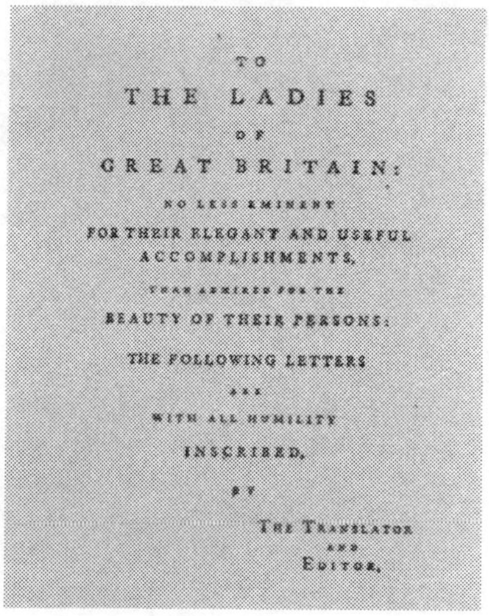

TO
THE LADIES
OF
GREAT BRITAIN:
NO LESS EMINENT
FOR THEIR ELEGANT AND USEFUL
ACCOMPLISHMENTS,
THAN ADMIRED FOR THE
BEAUTY OF THEIR PERSONS:
THE FOLLOWING LETTERS
ARE
WITH ALL HUMILITY
INSCRIBED,
BY
THE TRANSLATOR
AND
EDITOR.

Figure 33 Dedication in Thomas Martyn's translation of Rousseau's *Letters on the Elements of Botany*, 1785.

In outlining Thomas Martyn's academic career up to his College Fellowship and Tutorship, there has been no cause to mention his interest in botany. It is, of course, typical of the period that an academic education could still contain very little indeed that we would recognise as science, even if the student were strongly inclined towards a study of the natural world. As we have seen, medicine provided the only opportunity to study any science, and Cambridge had no very strong medical school. Thomas Martyn's botany had been acquired in a gentlemanly manner as a hobby of his father's, and this 'gifted amateur' status of botany is highly significant as a factor in the botanical tradition right up to the present day. Indeed, Martyn himself saw his botanical interest exactly in this light. In the Preface to the third edition of his *Language of Botany*, published in 1807, he wrote: 'Being then (1753–6) engaged in academical studies, and afterwards (1756–9) in those of the profession I had determined to adopt, Botany was rather the amusement of my leisure hours, than my serious pursuit.'

One thing in particular had occupied Thomas Martyn in his botanical hobby – the writings of Linnaeus which, as we have seen, his father was receiving from the author himself. 'About the year 1750', Martyn wrote in 1807, 'I was a pupil in the school of our great countryman, Ray. But the rich vein of knowledge, the profoundness and precision which I remarked everywhere in the *Philosophia Botanica* (published in 1751) withdrew me from my first master, and I became a decided convert to that system of Botany which has been since generally received.' In 1809 he wrote '. . . that inestimable work, the *Philosophia Botanica*, in 1751, and, above all, the *Species Plantarum* in 1753, which first introduced specific names, made me a Linnaean completely'.

The adoption of the Linnaean system was important, as Thomas Martyn obviously appreciated, because the 'binomial' specific name was an enormous improvement over the cumbersome, unstandardised Latin 'name-phrases' which were in use before Linnaeus. Martyn's first published work, *Plantae Cantabrigienses*, which appeared in 1763, was arranged strictly according to the Linnaean system and nomenclature; by means of this book, he began to teach about the local flora, using what are in most cases the names we still use today. The book, in spite of its title, is in English throughout, and consists of what we would now call a 'check-list' of the 829 species of plants known to occur in the County of Cambridgeshire, 'disposed', as the author explains, 'according to the method of the celebrated Linnaeus'. The list is in three columns, the accepted Linnaean binomials taken from Hudson's *Flora Anglica* (which had appeared in 1762) being in the first column, the synonym in John Martyn's *Methodus* in the second column, and that in Ray's *Catalogus* in the third.

It is interesting that Martyn adopted this check-list form, relating it to the new national Flora of Hudson. In the Preface he states quite clearly why: 'The publication of Mr Hudson's *Flora Anglica* has made it unnecessary to do more; I have therefore contracted my work into as narrow a compass as possible, chusing rather to refer the Cambridge botanist to him for the generic and specific characters of plants, than to transcribe them over again.' It is, in fact, a recognisably modern County Flora with its species arranged systematically in an internationally-accepted order, and moreover, with a special section, the *Herbationes Cantabrigienses*, which tells you just what species you might expect to find on thirteen selected 'journeys' or 'herborisations'. The first eight of these are 'within the compass of a moderate walk' and include all our standard excursion sites of the present day such as Madingley Wood, Grantchester, the Gogmagog Hills, and Cherry Hinton. The last-named herborisation is subdivided into several lists, of which the most precisely localised is that entitled 'in the close called by Mr Ray the Chalk-pit Close'; here are given fifty-nine species, most of which still grow there, including the famous Wild Cherry, *Prunus avium*, the Perennial Flax, *Linum perenne*, and the rare Perfoliate Honeysuckle, *Lonicera caprifolium*, 'found there by Mr Lyons' (Israel Lyons, who published a *Fasciculus* of Cambridgeshire plants in the same year, 1763). The tradition of 'herborising', established by Ray, appears now as a formal part of Thomas Martyn's teaching; it is a tradition which has continued (not without breaks, it is true) right up to the present day, and so constitutes a stable element in the Cambridge botanical studies traceable back for more than three centuries. An odd feature of the *Plantae Cantabrigienses* is that Cambridgeshire plants occupy only 43 pages; the greater part of the book (another 70 pages) consists of an Appendix of 'lists of the more rare plants growing in many parts of England and Wales', put together from many sources both published and unpublished.

The other part of the Linnaean method, the so-called 'sexual system', was not destined to last, as we shall see in the next chapter. It is, however, important to remember that the frankly artificial system of grouping the Linnaean genera had great advantages for identification and teaching; the success of Linnaeus' writings throughout Europe was at least partly due to the fact that the sexual system was easy to understand and to use, and, in an age when exploration of new floras was proceeding rapidly, this was an enormous practical advantage. It is also important that the Linnaean system was easy for amateurs. Thomas Martyn, for example, was responsible for the translation of Rousseau's *Letters on the elements of botany. Addressed to a lady*, which appeared in 1785, with a companion volume of *Thirty-eight Plates . . .* in 1788. The great burst of general and local Floras which took place after the publication of

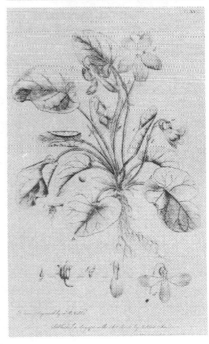

Figure 34   Illustration of *Viola odorata* from Thomas Martyn's *Thirty-eight Plates . . .*, 1788.

Hudson's *Flora Anglica* (Henrey calculates that about two and a half times as many appeared between 1763 and 1800 as there were from 1700 to 1762) can reasonably be attributed to the general convenience the Linnaean nomenclature and system afforded. In this movement to popularise botany on the Linnaean model, Thomas Martyn's lectures in Cambridge played their part. In a letter to James Edward Smith written at the age of 87 years, Martyn recalls his advocacy of Linnaean botany: 'his system can hardly be said to have been publicly known among us, till about the year 1762, when Hope taught it at Edinburgh, and I at Cambridge, and Hudson published his Flora'.

Martyn had intended to make a new, enlarged edition of his *Plantae Cantabrigienses* but, finding that Richard Relhan, Fellow and Chaplain of King's College, intended to write a Cambridgeshire Flora, he passed over to Relhan all his notes. Relhan's *Flora*, published in 1785, with appendices in 1786, 1788 and 1793, is chiefly remarkable for the useful treatment of the lower plants, the knowledge of which was very limited in the eighteenth century. In the Botanic Garden archives we have a letter from Thomas Martyn to his friend Pulteney, written from Little Marlow on 30 July 1783, in which Martyn explains his interest in Relhan's *Flora*, and says: 'Will it not be remarkable if Cambridgeshire should have four Floras published, before Oxfordshire has one?' This is indeed what happened, because Sibthorp's *Flora Oxoniensis*, the second British County Flora, was not published until 1794 – by which time Cambridgeshire had its succession of four Floras. Oxford-Cambridge rivalry, like many other Oxbridge traditions, is clearly long-established!

Election to the Royal Society came for Martyn in 1786, and some years later he became a Fellow of the newly-formed Linnean Society. His later botanical work was mainly the laborious compilation of a new four-volume edition of Miller's famous *Gardener's Dictionary*, on which he worked over 22 years, and which finally appeared in 1807. He died in Pertonhall in 1825, at the age of 89 years.

A complete bibliography of John and Thomas Martyn's works has been compiled by Albu (1956).

The 'Old Botanic Garden'

We come now to the successful foundation of the Botanic Garden, on a plot of ground purchased by the Vice-Master of Trinity College, Richard Walker, on 16 July 1760, and given to the university on 25 August 1762 in trust 'for the purpose of a public Botanic Garden'. How much either of the Martyns was involved in the creation of the new Garden is difficult to decide from the available texts. If we are to judge from a very diplomatically-worded reference to John Martyn in the anonymous pamphlet: *A short account of the late Donation of a Botanic Garden to the University of Cambridge* (actually written by Walker himself), we might well conclude that the then Professor was very little involved. Walker says: 'We have generally had titular Professors of Botany, but nothing worth mentioning left behind them: Dr Martyn indeed within our memory, laboured much to bring this science into repute; read public lectures for several years; perambulated the country with his scholars, shewing them the Cambridgeshire plants where Mr Ray had described them to grow, and making many additions to that Catalogue: but this gentleman's private affairs took him from us, much esteemed for his great knowledge of plants.'

So far as Thomas Martyn is concerned, we must assume that Richard Walker's generous donation was one of the factors which made his election to the Chair of Botany the more welcome. Neither he nor Walker, however, seems to have assumed that the Chair itself would bring any obligation to lecture in Botany, and we find that the establishment of the Garden provides for 'two officers, a Reader on Plants, and a Curator or Superintendent of the works in the Garden'; further that 'the said Vice-Master has appointed the Reverend Mr Thomas Martyn, now Titular Professor of Botany, to be the first Reader; and Mr Charles Miller, the first Curator'.

Walker states quite clearly what the Garden is for, referring particularly to the fact that 'about fifteen years ago, the learned physician, Dr Heberden, was so kind as to oblige the University with a course of experiments, upon such plants as he then found among us, in order to show their uses in medicine . . . But this Doctor's great abilities in his profession soon after called him from us, much lamenting the want of a Public Garden, furnished with sufficient variety of plants for making the like experiments.' Heberden's *Materia Medica* cabinet, with all the materials, mineral, plant and animal, is preserved in the Library of his College, St John's. It is accompanied by the heads of the course of 31 lectures which began on 7 April and finished on 22 May 1747 . . . with a week off between the tenth and eleventh lecture for the Newmarket races!

The purpose of the new Garden is clearly stated in the first of the 'Statutes or Orders' established by 'the Vice-Master with approbation of the Trustees'. It runs as follows: '*Imprimis*. That as soon as the stoves and greenhouse are finished for the reception of tender plants, and the whole Garden perfected for the hardy sorts, trials and experiments shall be regularly made and repeated, in order to

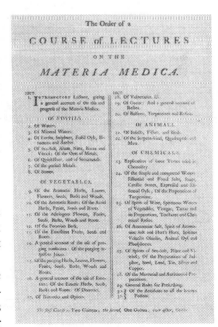

Figure 35   Heberden's *Course of Lectures*, 1747.

Figure 36   View of the main entrance to the Old Botanic Garden in the time of
Thomas Martyn.

discover their virtues, for the benefit of mankind – This the said
Vice-Master declares to be the principal interest of the Garden: in
comparison whereof, Flowers and Fruits must be looked upon as
amusements only; though as these do not want their excellencies
and uses, they need not be totally neglected.'

This is a fascinating, if depressing, passage, explicitly rejecting
both pure botany, and agricultural or horticultural research or
teaching, as main purposes of the new Garden. Bradley's three
purposes for *his* Garden only some thirty years earlier have been
narrowed down to the single, respectable and traditional one, re-
establishing the dominance of medicine over botany.

Perhaps this limitation would not have been so serious if Cambridge had had a really important medical school, and Thomas
Martyn had been a man with the ability and breadth of interest of
his friend John Hope, who became at almost the same time (1760)
Professor of Botany and Materia Medica in Edinburgh. (An excellent account of Hope's many-sided contribution to botanical
teaching is given in Fletcher & Brown (1970).) As it was, we get the
impression that the new foundation, though a great improvement
on the previous state, was never a really flourishing concern, and
indeed was continually faced with financial and other difficulties
which became chronically depressing after Thomas Martyn retired
from the university scene altogether in 1796.

Two unfortunate happenings made the going much harder for the new Garden. The first was the death of Richard Walker himself in 1764, which removed early from the scene the donor who had both influence and enthusiasm to devote to the development of the Garden. The second was the departure of the first Curator, Charles Miller (son of Philip Miller, of the Chelsea Physic Garden) in 1770 for the East Indies, after only seven years' work in establishing the Garden.

How far Miller's resignation was due to the inability of the Trustees to pay him a tolerable salary is not clear, but financial difficulties certainly made Thomas Martyn take on the Curatorship himself on an unpaid basis for some years after Miller's resignation. Eventually the Curator's post was again filled, but, from the series of holders, the only name which means much today is that of James Donn, who held the post from 1794 to his death in 1813, and published the *Hortus Cantabrigiensis*, a check-list of plants grown in the Garden. This work became so famous that it deserves more detailed comment and some comparison with the earlier Catalogues of the Walkerian Garden (see p. 45).

Reading some of Thomas Martyn's letters to his friends, it is easy to feel sorry for him, for botany was still a 'Cinderella' of subjects. To Richard Pulteney he writes in April 1761: 'Our Garden begins to flourish, shrubs and trees are already planted; plenty of seeds, both tender and hardy, are sown; a stove is building; and Stone is preparing to raise the superstructure of a greenhouse on the foundation which was laid last year. All this, I hope, will increase the number of botanists among us. Indeed, we already begin to grow considerable, for I never had more than one companion before this spring, but now I have three; and expect soon to have two or three more converts.'

The land on which the Garden was laid out was the Mansion House in Free School Lane, on the site of the ancient Monastery of the Austin Friars, 'with near five acres of garden about it, well walled round, quite open to the south, conveniently sheltered by the Town on the other quarters, with an antient watercourse through the midst of it'. No doubt it seemed an adequate site for a traditional 'physic garden', comparable in size to those at Oxford and Chelsea, and Martyn's botany lectures seem to have started well there in 1763, when fifty students attended. These lectures were published in outline, together with a short preface and a dedication to Richard Walker, as a small pamphlet in 1764. The preface is interesting in that the case for studying botany made here by Martyn is entirely in terms of the newly-discovered exotic floras of the world, and the great value which knowledge of them might bring to the country and the empire. He refers specifically to the 'great spirit of planting which has lately arisen in Britain, and the noble taste which now prevails in gardening', which 'give room to hope that botany at least may be pursued among us with more ardour as a science'. He pays tribute to Edinburgh botany under Hope, Chelsea under Miller, and 'the noble garden at Kew . . . excellently furnished and considering how few years it has subsisted, . . . in wonderful forwardness'. There is no word about teaching the medicinal uses of plants. In 1766, however, he writes: 'the Garden gets on very well in point of plants under the direction of Mr Miller, but our income is still very scanty, so that we cannot finish our greenhouse, much less build stoves; indeed, we are obliged to use a degree of frugality not very consistent with the dignity of an University, or the usefulness of the design, but we keep it on foot for better times'. By 1770, as we have seen, he was obliged to take over the job of Curator himself.

In 1771 Martyn published the *Catalogus Horti Botanici Cantabrigiensis,* which contains as a Preface the heads of his course of lectures, and sets out in Linnaean order the plants in cultivation. A small *Mantissa* (Supplement) of 31 pages published in the following year includes a plan of the Garden: its main features include the Systematic Beds arranged according to the Sexual System and a range of glasshouses against the north wall. Space for trees and shrubs is very limited; these are planted round the periphery of the Garden. The main gravel walk runs north–south and crosses a rectangular pond running east–west. The fine wrought-iron gates in Pembroke Street (now re-erected as the Trumpington Street gates of the present Garden) are not visible on this plan; they were apparently erected a little later, though the exact date is not known. The most familiar picture of the Old Garden is that from Ackermann's History of Cambridge, 1815. It shows a tasteful, decorous scene, but the detail, and in particular the individual exotic-looking trees, must owe something to the artist's imagination.

The old Mansion House, which Walker had bought with the land, and in which the ground-floor rooms were for the Professor and Walkerian Reader to teach and lecture, was sold in 1784 to Mr John Mortlock (of Mortlock's, now Barclays Bank, on the present site), and the University erected in 1787 in the Garden a lecture room and ancillary buildings to be used, originally by the Professors of Botany and the Jacksonian Professor of Natural Experimental Philosophy, and later also by the Professors of Chemistry, Anatomy and Physics. In this way the 'New Museums Site' gradually became the centre of natural science in the University, with what further developments in the nineteenth and twentieth centuries we can now appreciate. To tell the story of the final years of the Old Garden we must, however, wait for Henslow to appear.

Figure 37   Ackermann print of the Old Botanic Garden, 1815.

Note on The Collections in the Walkerian Garden

Thomas Martyn's Catalogue (1771), produced, as we have seen, when he took over the Curatorship, is obviously a genuine list of plants in cultivation ten years after the Walkerian Garden was opened. It includes, for example, no Sedges (*Carex*), and only three Willows (*Salix*), although many species of both these genera were known, and indeed available, at the time. It is essentially a *teaching* collection, put together by Charles Miller during the time when the Professor was actively lecturing and using the plant material in his lectures. Moreover, one of the copies of the book still preserved in the Cory Library of the Botanic Garden is interleaved, and has many notes, presumably by Martyn himself, of plants newly acquired for the collection – examples are sugar cane (*Saccharum*) and the Madagascar periwinkle (*Vinca rosea*, now called *Catharanthus roseus*).

A small, anonymous Catalogue of the Garden, published in 1794, is probably by Philip Salton, the Curator of the time, who was succeeded at the end of that year by James Donn. Donn was a very able botanist and gardener, and had been trained under Aiton at Kew; he looked after the Garden until his death in 1813, and published seven editions of his famous *Hortus Cantabrigiensis* between 1796 and 1812. Those who have considered Donn's *Hortus* have generally agreed that, whilst the later editions (especially the posthumous ones, which appeared up to 1846) are obviously lists of all plants known in cultivation in gardens in England in general and do not refer to the Cambridge collections only, the earliest editions could be a reliable guide to the content of the Cambridge Garden. This seems improbable, for Donn himself, in the Preface to the first edition (1796), implies that he has not so limited his selection: '[the Catalogue] is intended for the use of those students in botany who shall be disposed to inspect the productions of the Walkerian Garden. From it they will immediately learn what plants they may have an opportunity of finding there, *and what are yet required to render the collection more worthy of their notice*'. (My italics). A comparison between Martyn's Catalogue and that of Donn would seem to bear this out. Thus, Donn includes sixteen willows, and no fewer than 40 species of *Carex*, some of them bog and acid moorland species which would be quite difficult to grow.

To say this about Donn's *Hortus* is in no way to belittle his knowledge or his ability to build up and maintain an impressive collection of plants using his links with Kew. We have independent witnesses to the size and interest of the collections during Donn's Curatorship. Thus, an entomologist, William Kirby (Freeman

Figure 38   Herbarium specimen of *Silene maritima* made by Donn, presumably from a plant cultivated in the Old Botanic Garden.

1852), recorded in his journal his impressions of a visit paid to the Garden in July 1797: 'Mr Newton of Jesus accompanied us to the Botanical Garden, which, by the abilities and industry of Mr Don (*sic*), the Curator, is now in excellent order; the collection of plants is greatly augmented, and the labels are in general accurate'.

This standard seems to have been maintained to some extent by the next Curator, Arthur Biggs, for we find that Schultes, a celebrated Austrian botanist, who visited the Garden as late as 1826, has left us an account of his mixed impressions. Although he starts by saying that it 'contains about five acres of very bad ground', he is nevertheless favourably impressed by the neat, well-arranged appearance, and the size of the collections ('from five to six thousand species'), particularly the 'Alpine plants, among which are some rare species from the Scotch Highlands . . . very properly cultivated in small pots and placed during winter under glass'. It seems that Donn's influence survived to the end of the Old Garden, and it is fascinating to find that 'alpines' were a special feature of the Cambridge Botanic Garden more than 150 years ago, as they still are today.

5

Henslow and the rise of natural science

Figure 39 *Dipsacus strigosus* in the Churchyard of Little St Mary's Church, Cambridge, where Henslow was curate in 1824.

By the end of the eighteenth century the teaching of botany in the University had once again lapsed, with Thomas Martyn an aged absentee Professor; however, as we have seen in the previous chapter, the Walkerian Garden was probably at its best at this time under the efficient, devoted Curatorship of James Donn. After Donn's death the Garden also declined, sharing in the general neglect of the science. It is true that in 1819 there was a move to appoint James Edward Smith as Deputy Walkerian Reader in the Garden so that a course of lectures in botany might be re-started, but this came to nothing because of opposition to Smith as a Dissenter. (The incident and its aftermath produced an exchange of publications, of which Smith's pamphlet (1818) entitled *Considerations respecting Cambridge, more particularly relating to its Botanical Professorship* is well worth reading.)

The death of Thomas Martyn in 1825 made vacant the Botanical Chair or, more strictly, two legally separate Botanical Chairs which Martyn had held – the University Professorship, created originally for Bradley, and the Regius Professorship made for Martyn in 1793 and to which a stipend of £200 a year was attached. At the time there was a complex political battle within the University affecting the right of appointment to Chairs (see Winstanley 1940, pp. 29–41 for a full explanation) and, as a result of compromise, only the new Regius Professorship was filled, the older University Chair being left in abeyance. The successful candidate was a young man of 29 years, John Stevens Henslow, who was already Professor of Mineralogy. His election was to lead to a rejuvenation of botany in Cambridge.

Henslow was born in 1796 in Rochester, Kent, where his father, originally a solicitor, had settled down as a wine-merchant and later married into the brewery business. John was the eldest of eleven children, whose early curiosity for natural objects seems to have been stimulated and encouraged by both his parents. In 1805 he was sent to a private school in Camberwell, where he found the drawing master was a keen entomologist, and where he developed butterfly-collecting as a hobby. From these beginnings began an acquaintance with professional entomologists at the British Museum and, indeed, an ambition to become an explorer and collector of animals, which he eventually reluctantly abandoned through parental opposition. The young Henslow seems to have found his school studies easy, and he was prepared to come to Cambridge, entering St John's College in October 1814. In Cambridge he found there were university lectures in chemistry by Cumming, and in mineralogy by Clarke, and took both these courses enthusiastically, though neither was relevant to his Tripos studies, which had to be principally in mathematics. He got a tolerably good B.A. degree at the end of three years.

The Professor of Mineralogy, E. D. Clarke, was a brilliant, enthusiastic lecturer who had travelled widely in Europe and Asia, collecting not only rock specimens, but also manuscripts, paintings and other *objets d'art*. He must have had a profound influence on the young Henslow. Winstanley (1940) says of Clarke: 'He seems to have enjoyed his lectures quite as much as his hearers, whom he carried away on the tide of his own enthusiasm. He made a practice of speaking without notes . . . If he had been a popular entertainer, he could hardly have taken greater pains to amuse; and about the year 1816 he began "to study oil painting, for no other purpose than to embellish his lecture room with fresh ornaments and attractions, and by a series of designs to give a faithful and accurate representation of the native character and situation of his most remarkable minerals, and of the scenes amidst which they occur".'

Cumming's lectures in chemistry must have been more prosaic, but Henslow also took up the study of that science enthusiastically, and Jenyns (Henslow's biographer and brother-in-law: see below) records in his *Memoir* (1862) that, 'when I attended Professor Cumming's lectures in 1820, Henslow assisted him in the lecture-room, by which he must have gained much practical knowledge of that branch of science'. Jenyns continues: 'Zoology, however, seemed then to be his favourite pursuit . . . Botany had scarcely been taken up. . . .' Chronologically the third Cambridge scientist to influence Henslow was the geologist Adam Sedgwick, who became Woodwardian Professor of Geology in 1818, and took Henslow with him during the Easter vacation of the following year on a geological tour of the Isle of Wight. This was so successful an initiation into field geology that Henslow himself undertook a geological survey of the Isle of Man in the following Long Vacation, taking with him a group of Cambridge students as his pupils, and publishing his results in the *Transactions of the Geological Society* in 1821 – his first learned paper, produced when he was 25 years of age.

Figure 40   Lecture-rooms in the Old Botanic Garden, as Henslow knew them.
The engraving is dated 1800.

The reader might well ask at this point why Henslow ever
applied for the Chair of Botany, and even more why his application
was successful. It is not entirely clear how we should answer either
question. If we are to believe Jenyns: 'the Professorship of Botany
was the one to which he had been looking for some years, and for
which he had been preparing himself at a time when he never
anticipated that the Chair of Mineralogy would be open to his
acceptance first'. This does not easily square with Jenyns' own
assessment of Henslow's interest in the subject in 1820, quoted
above. Perhaps we should answer the first question by saying that
Henslow's field enthusiasms were strongest in zoology, and that he
accepted botany as a second best in the complete absence of any
formal recognition of zoology (as distinct from medicine, of
course). It seems, however, that his switch from the Chair of
Mineralogy to that of Botany might also have been motivated by
the feeling that, in succeeding the brilliant 'showman-lecturer'
Clarke, he found it impossible to hold such large, enthusiastic
audiences in mineralogy, and saw in the neglected science of botany
a great opportunity. As Henslow himself later admitted (1846):
'when appointed, I knew very little indeed about botany . . .' but,
he added, 'I probably knew as much of the subject as any other
resident in Cambridge'. So much for Henslow's motives, but what
of the electors to the Chair of Botany? Here we can only guess but,
as we have seen, much political activity surrounded the appoint-

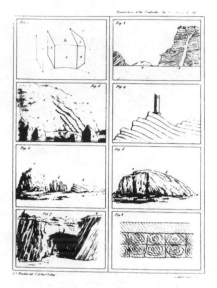

Fig. 1. Cleavages exhibited by the strata of the quartz rock . . . p. 364

Fig. 2. Vertical section of a mass of breccia (a), and a quartzose vein (b) connected with it, which rises through the chlorite schist, near its junction with the quartz rock .... p. 366

Fig. 3. Junction of the quartz rock (a), and chlorite schist (b) to the West of Rhoscolyn .... p. 366

Fig. 4. Section of the stratified chlorite schist .... p. 371

Fig. 5. Serpentine (a) rising abruptly through the chlorite schist (b), which dips in various directions .... p. 376

Fig. 6. Massive serpentine (a) gradually assuming a schistose character (b) .... p. 376

Fig. 7. Appearance presented by the greywacké slate on the shore near Monachdy .... p. 383

(a) Hard, green, and unlaminated portion, passing gradually on one side to a schistose black slate (b), and terminated abruptly against a similar rock on the other.

Fig. 8. Arrangement of particles in the stratified grit at Bodorgan, p. 395.

Figure 41 Illustrations from Henslow's early paper on the geology of Anglesey, 1821, showing his talent for sketching.

ments to the two University Chairs, and Henslow's election was certainly not free from complication of that sort.

From the earliest days of his graduate career in Cambridge, Henslow showed a strong desire to share his enthusiasm for natural history, a desire which gradually, as his influence grew, turned into a general concern for the teaching of natural science in the University. During his first field trip, with Sedgwick in the Isle of Wight, he had discussed with the Professor a scheme to found a 'corresponding society' to promote the study of science in Cambridge. This was the beginning of the Cambridge Philosophical Society, which was duly founded in November 1819, and held its first meeting in the Museum of the Botanic Garden in December of the same year. As a junior member of the University, Henslow was presumably considered too young to hold office immediately, but he became joint Secretary in 1821 and continued as Secretary until 1839, when he ceased to reside in Cambridge. His *Geological Description of Anglesea*, published soon after his first paper already mentioned, appeared in the first volume of the *Transactions of the Cambridge Philosophical Society* in 1821, and his most interesting botanical paper: *On the examination of a hybrid Digitalis*, appeared ten years later in the same journal (Walters, S. M. ed. 1981).

One of the undergraduates obviously influenced by Henslow was Leonard Jenyns, of Bottisham Hall, who records that they first met in 1820, and who shared with Henslow a passion for local zoology. Through the friendship with Jenyns Henslow met his future wife, Harriet, Jenyns' sister; they were married on 16 December 1823 and set up house in Cambridge. As Professor of Mineralogy Henslow had already established a reputation in the University as a scientist, and he now began the practice of regular Friday evening soirées to which he invited anyone interested in science, and especially in natural history. The most impressive tribute to the influence of Henslow's lectures, excursions and soirées comes from his most famous pupil, Charles Darwin, who wrote (Jenyns 1862): 'His lectures on Botany [this was 1828, when Darwin arrived in Cambridge] were universally popular, and as clear as daylight. So popular were they, that several of the older members of the University attended successive courses. Once a week he kept open house in the evening, and all who cared for natural history attended these parties, [where] . . . I have listened to the great men of those days, conversing on all sorts of subjects, with the most varied and brilliant powers. This was no small advantage to some of the younger men, as it stimulated their mental activity and ambition. Two or three times in each session he took excursions with his botanical class . . . He used to pause every now and then and lecture on some plant or other object; and something he could tell us on every insect, shell, or fossil collected, for he had attended to every branch of natural history. After our day's work we used to dine at some inn or house, and most jovial we then were.' The friendship between Darwin and Henslow deepened and became permanent, and Henslow's role in recommending Darwin for the post of naturalist on the famous voyage of the *Beagle* was crucial in the development of the young naturalist's ideas on evolution by natural selection.

The description given by Jenyns of Henslow's lectures is quite fascinating, especially in that it reveals how far Henslow was responsible for several innovations which have survived, even flourished, to the present day. Most importantly, he invented the practical class and its accompanying 'Demonstration Bench'. Listen to Jenyns' first-hand description:

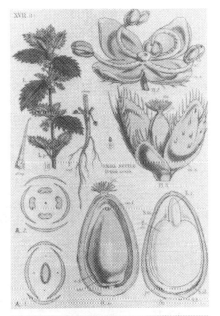

Professor Henslow had already had three years' experience in lecturing, and he neglected nothing in his power to make his lectures attractive and popular, without departing from a plan of study that could alone give the students a proper grounding in Botany as a science. One great assistance he derived from his admirable skill in drawing. His illustrations and diagrams . . . representing all the essential parts of plants characteristic of their structure and affinities, many of them highly coloured, were on such a scale that when stuck up they could be plainly seen from every part of the lecture-room. He used also to have 'demonstrations', as he called them, from living specimens. For this purpose he would provide the day before a large number of specimens of some of the more common plants, such as the primrose, and other species easily obtained, and in flower at that season of the year, which the pupils, following their teacher during his explanation of their several parts, pulled to pieces for themselves. These living plants were placed in baskets on a side-table in the lecture-room, with a number of wooden plates and other requisites for dissecting them after a rough fashion, each student providing himself with what he wanted before taking his seat. In addition to these were rows of small stone bottles containing specimens of all the British plants that could be procured in flower, the whole representing, as far as practicable, the different natural families properly named and arranged.

Henslow's own advice to his students, from the Preface to his *Questions on the subject-matter of Sixteen Lectures in Botany* (1851) is beautifully clear, and could be repeated with advantage today.

Figure 42 Two of Henslow's 'Botanical Diagrams', prepared for teaching plant families and published (in colour) in 1854.

Whoever may be expecting to acquire a competent knowledge of this subject by merely listening to what shall be told him at lectures, will be disappointed. "How to observe" is an art to be acquired by "observing" and not by listening, or even by reading alone. The Student will find himself confused rather than enlightened if he will not take the trouble to examine plants, and to compare what he sees in them with the descriptions and definitions by which they are to be recognised. If he will consent to do this, he will soon find a growing interest in the subject: and it is certainly one which need not interfere with the regular course of reading exacted of him for his degree. My advice to all who desire to master as much of Botany as may be required for a Pass-examination, is to attend these Lectures in their first year. In occasional visits to the Botanic Garden, during walks in the country, and especially in the long vacation, ample opportunities will be found for acquiring far more than will be necessary for this object.

That Henslow's name should be known to future generations only as the teacher of Charles Darwin would in no way have disappointed him, for he seems to have been universally liked and admired as a modest, principled man who sought no fame for himself. Again, to quote Darwin: 'He never took an ill-natured view of anyone's character, though very far from blind to the foibles of others. It always struck me that his mind could not be even touched by any paltry feeling of vanity, envy or jealousy.'

One other claim to fame illustrates a different side of Henslow's character. In 1826, together with the Master of Corpus Christi College, the Rev. J. Lamb, he organised a successful opposition to the practice of bribing (by paying the expenses) of non-resident voters in university elections (Henslow & Lamb 1826), and later took up a similar concern about bribery and intimidation by the Tory supporters in the Borough election, stating in a published *Address to the Reformers of the Town of Cambridge* (1834) that he felt it his duty 'as a clergyman' to offer 'a strenuous resistance to every kind of appeal . . . to the weakness and corruptions of our human nature'. In the following year he took the bold step of agreeing to act as 'common informer' to bring to Court the offending Conservative agents, who had again resorted to bribery; by doing this he incurred abuse and persecution, and the slogan 'Henslow, common informer' was written as *graffiti* around the streets of Cambridge. One of these slogans appeared in black capital letters on the face of the then newly-built Wilkins building of Corpus, near St. Botolph's Church, and lasted, faint but legible, for about 100 years; the last traces were finally removed when the building was cleaned in the early 1960s. (A photograph of this slogan taken in 1955 appears in Jean Russell-Gebbett's biography of Henslow, published in 1977.)

How far the idea of ordination and the Church as his eventual career was present in Henslow's mind from his early days is difficult to assess. Jenyns says that his parents certainly wanted him to go into the Church: but Jenyns also hints that he would gladly have remained an academic, contributing to the growth of scientific study in the University, had it been financially possible for him to do so. Henslow's own writings throw little light on his motives, except for one revealing passage – incidentally, one of the very few places where he ever complained of any difficulties in his life – written in 1846, where he recalls that in early years he had to supplement his meagre income by giving private coaching to undergraduates (not, of course, in his beloved natural science, which still played no part in the Tripos). In this passage he observes that 'five or six hours a day devoted to cramming men for their degrees is so far apt to weary the mind as to indispose it for laborious study, especially if we happen not to be fitted with talents or energy sufficient to overcome such an obstacle'. Whatever the mixture of motives, Henslow was ordained, and became a Curate at Little St. Mary's Church, both in the same year 1824. In 1832, he obtained the living of Cholsey-cum-Moulsford in Berkshire, and spent the Long Vacations there. He remained in Cambridge until 1839, two years after being presented by the Crown to the valuable living of Hitcham, in Suffolk, which brought him an income of over £1,000 a year, and presumably solved all his financial problems. Jenyns says: 'he did not immediately cease to reside in Cambridge . . . but in 1839, finding that the duties of so large a parish could not be properly attended to except by living constantly amongst his people, he came to the determination to quit the University altogether . . .'. Although, as we shall see, Henslow continued to play some part in University affairs (and particularly in the development of his 'New Botanic Garden'), his great influence as a teacher in Cambridge was at an end, and the parishioners of Hitcham, and school education in England as a whole, increasingly benefited from his talents.

When appointed to the University Professorship of Botany in 1825, Henslow was at the same time made Walkerian Reader by the Trustees of the Botanic Garden. Since the Chair itself provided no statutory duties, his first area of practical concern was, as Walkerian Reader, to rescue what he could of the material accumulated by his predecessors in the Botanical Museum, which was in a state of gross neglect. Henslow set about his work with characteristic vigour. He found the Martyn Herbarium badly damaged, but rescued what he could (probably about one-tenth of the original collections), and by doing so laid the foundation of the important Herbarium which the University still possesses. The Martyn collections which have survived contain a good many specimens collected in the earliest days of the Walkerian Garden, in addition to even earlier historical collections such as those made by Houstoun in Mexico and Jamaica and sent to John Martyn. So far as the Garden itself is concerned,

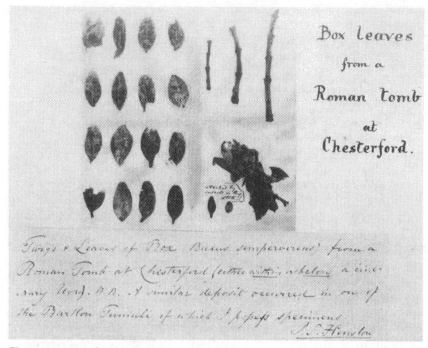

Figure 43   Henslow's specimen in the Cambridge University Herbarium of Box leaves from a Roman tomb.

Henslow must have soon decided that it really had no future on the restricted original site; as he himself put it, the Garden was 'utterly unsuited to the demands of modern science'. His representations to the Governors, which must have been repeated at regular intervals, bore fruit in 1831, when a special Act of Parliament authorised the purchase by the University from Trinity Hall of the present site on which a new Botanic Garden was planned.

Before we look at the provisions of the Act, and Henslow's plans for the 'New Garden', we might note a peculiar hazard attendant upon the attempts to keep a scientific plant collection on the old Walkerian site. In *Loudon's Magazine for Natural History* (anon. 1833) appeared the following fascinating 'nature note':

Jackdaws are comparatively numerous at Cambridge. The Botanic Garden there has three of its four sides enclosed by thickly built parts of the town and has five parish churches and five colleges within a short flight of it. The jackdaws inhabiting (at least for a certain time in each year) these and other churches and colleges had, in the years 1815 to 1818 . . . discovered that the wooden labels placed before the plants whose names they bore in the Botanic Garden would well enough serve the same purpose as twiggy sticks off trees, and that they had the greater convenience of being prepared ready for their use and placed very near home. A large proportion of the labels used in this Garden were made out of deal laths and were about nine inches long, and about an inch or more broad, . . . and although of this size, as they were very thin when dry, pretty light. To these the jackdaws would help themselves freely whenever they could do so without molestation, and the time at which they could do this was early in the morning before the gardeners commenced work for the day, and while they were absent from the Garden at their meals, and the jackdaws would sometimes fetch away labels during the gardeners' working hours from one part of the garden, when they observed the gardeners occupied in another, as was often the case in their attending to the plants in the greenhouse, etc. . . . Those who are aware how closely some species of the grasses, garlic, umbelliferous plants, etc., resemble each other and also how needful it is to prefix labels to them, as remembrances of their names, will readily perceive that much inconvenience arose from the jackdaws appropriating some of the labels; and this especially when they removed, as they sometimes did, the labels from sown seeds, as the plants arising from those seeds must in some species grow for a year or more before their names could be ascertained. I cannot give a probable idea of the number of labels which the jackdaws annually removed, but have been more than once been told by persons who had ascended the tower of Great St Mary's Church and the towers or steeples of other churches that wooden labels bearing botanical inscriptions were abounding in these places. The house of the late Dr Kerrick, in Freeschool Lane, was close beside the Botanic Garden; and the shaft of one of the chimneys of his house was stopped up below, or otherwise rendered a fit place of resort for jackdaws. From this chimney shaft Dr Kerrick's man-servant got out on one occasion eighteen dozen of the said deal labels; and these he brought to Mr Arthur Biggs, the Curator of the Botanic Garden. I saw them delivered and received . . . This number of labels and the fact of the occurrence of plant labels on other buildings about the town prove that in general terms the aggregate of labels lost from time to time could not be inconsiderable.

The Act, dated 30 March 1831, authorises an exchange of land between the University and Trinity Hall, and 'the removal of the present Botanic Garden . . . to a new and more eligible site'. The land acquired by the University was 38 acres and 23 perches in extent – essentially the present Botanic Garden site – and the plot of land acquired by Trinity Hall in part exchange and measuring rather more than seven acres is the land on which Bateman St and Norwich St now stand. The difference in value, paid by the University to Trinity Hall, was £2,210 8s.

That a simple transfer of land should require an Act of Parliament 22 pages long may come as a surprise to many readers, but it was necessary because all sales of land by Colleges, until the middle of the nineteenth century, could only be done on authority of a Private

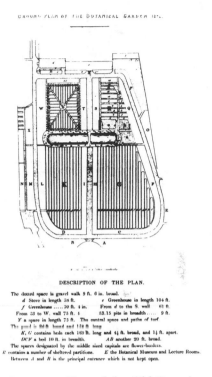

DESCRIPTION OF THE PLAN.

Figure 44   Plan of the Old Botanic Garden showing the formal arrangement characteristic of 'Physic Gardens'.

Figure 45   Site of the present Botanic Garden in 1809: view from Trumpington Road at the milestone at the corner of what is now Brooklands Avenue.

Act of Parliament. The Act sets out the argument (presumably from a draft supplied by Henslow, though the wording is heavily legal) in favour of the removal of the Botanic Garden from its present, inadequate site: 'although the site of the said Botanic Garden was at the time of the . . . indenture of release and assignment near the outskirts of the town of Cambridge, yet by the great increase and extension of the same town the said Botanic Garden is now clearly surrounded by buildings, whereby the free circulation of air is impeded and obstructed . . .' This passage concludes: 'it is therefore desirable that a new Garden should be formed of larger extent, and in a more open, commodious and eligible situation'.

The clearest and most accessible account by Henslow of his reasons for persuading the University to make a new and larger Botanic Garden are to be found in his *Address to the Members of the University of Cambridge on the expediency of improving, and on the funds required for remodelling and supporting, the Botanic Garden*. This twenty-page pamphlet, printed in 1846, contains a 'a few observations to the members of our University, inviting attention to what may be considered requisite for a modern Botanic Garden', and in developing this theme Henslow produces a case for botany as a whole. Before looking at this wider argument, let us see why Henslow thinks a larger Garden is necessary.

It is indeed true enough that one man with half-a-dozen flower-pots may do more towards advancing Botany than another will feel inclined to attempt with twenty or thirty acres of garden at his command: but it may very safely be asserted, that the larger the number of living species that are cultivated in a Botanic Garden, the greater will be the facilities afforded to us all; not merely for systematic improvement, but for anatomical and other experimental researches essential to the progress of general physiology. It is impossible to predict what particular species may safely be dispensed with in such establishments, without risking some loss of opportunity which that very species might have offered to a competent investigator, at the exact moment he most needed it. The reason why a modern Botanic Garden requires so much larger space than formerly, is chiefly owing to the vastly increased number of trees and shrubs that have been introduced within the last half century. The demands of modern science require as much attention to be paid to these, as to those herbaceous species which alone can form the staple of the collections in small establishments. The considerable portion of the ground which would be devoted to an Arboretum may be kept up at very much less expense than the rest, but would add very greatly to the ornamental as well as to the efficient character of the Garden.

Figure 46 Old label on Commemorative Lime Tree at entrance to Garden.

Two new ideas stand out here. Firstly, the provision of a range of living plants is seen as a service to the science of Botany as a whole, and medicine is no longer even mentioned as a justification. Secondly, the idea of an Arboretum appears for the first time so far as Cambridge is concerned, and is explicitly contrasted with the 'herb collections' of the traditional Physic Gardens and their direct descendants. The Arboretum, scientifically necessary to accommodate the newly-introduced woody plants from temperate Asia and North America in particular, would also have the further great advantages that it would be relatively cheap to maintain, and constitute an attractive amenity.

The contrast between the formal eighteenth-century Walkerian Garden and the 'New Garden' planned by Henslow is very striking indeed. Henslow's vision, though thwarted and restricted by circumstances, as we shall see, was an open, optimistic one which foresaw a rapid expansion of all the natural sciences, including a botany liberated from the systematic obsession. Henslow expresses it in this way:

But still I must consider the claims of Botany are not sufficiently appreciated among us. There are persons of great mathematical and classical attainments who have very erroneous notions respecting the ultimate aim and object of this science. Many persons, both within and without the Universities, suppose its objects limited to fixing names to a vast number of plants, and to describing and classing them under this or that particular 'system'. They are not aware that systematic Botany is now considered to be no more than a necessary stepping-stone to far more important departments of this science, which treat of questions of the utmost interest to the progress of human knowledge in certain other sciences which have been more generally admitted to be essential to the well-being of mankind. For instance, the most abstruse speculations on animal physiology are to be checked, enlarged, and guided by the study of vegetable physiology. Without continued advances in this latter department of Botany, the progress towards perfection in general physiology must be comparatively slow and uncertain. As regards the progress of Botanical physiology, even Chemistry itself must be viewed as a subordinate assistant, whilst it is making us acquainted with those physical forces by which mere brute matter is regulated and arranged.

An inspection of the contents of Henslow's lecture-course bears out this contrast of attitude. The *Sketch of a course of lectures in Botany for 1833*, for example, is divided into two parts, of which the first, headed *Demonstrative – on Tuesdays and Thursdays*, contains what we would call morphology, anatomy and systematic botany, and the second, headed *Physiological – on Mondays, Wednesdays and Fridays*, covers all 'traditional' plant physiology and includes flowering, fruiting and 'various modes of reproduction'. The 'Demonstrative Lectures', using common British plants as live material, are those described above by Jenyns: they combine an explanatory, illustrated lecture with a period of individual practical work for each student. How far Henslow's plant physiology is also experimental is not clear; perhaps he limited himself to 'demonstrations' of root pressure, osmosis, etc. which lend themselves to experimental presentation to the whole class.

The 1833 *Sketch* also includes recommended text-books in a list headed *Elementary Botanical Works* which begins with Lindley's *Introduction to Botany*, published the previous year. Henslow's own text-book, entitled *Descriptive and Physiological Botany*, obviously based on his lecture-course, was published in 1836, and, according to Jenyns, 'was long considered the best manual of structural and physiological botany in the English language'. To equip himself with up-to-date knowledge of plant physiology, he had studied de Candolle's *Physiologie Végétale* which appeared in 1832, and had indeed published a very able review of this important work in the same year.

Henslow's Easter Term course was apparently given without break for twenty-five years between 1825 and 1850, although the attendance seems to have been relatively poor in the later years. This can obviously in part be put down to the fact that he was no longer living in Cambridge; in 1846 we find his pupil and eventual successor in the Chair of Botany, Babington, complaining in a letter to Professor Balfour: 'Never was botany at so low an ebb as now in this place. A non-resident Professor, who only comes here

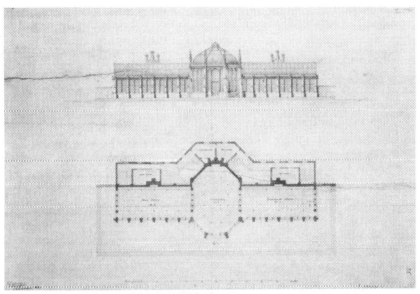

Figure 47   Lapidge's design for the glasshouse range dated 1830. This design was never carried out; the eventual range was much more modest.

for five weeks (as he calls it), going away on Saturday morning in each week and returning Monday evening.' If the loyal Jenyns is to be believed, however, the attendance in the 1840s was probably at its lowest due 'to the additions made about that time to the examination for a common degree, which left men less leisure for studying natural history'. Apparently after 1852 the numbers again increased, because some attendance at the lectures was made compulsory for ordinary-degree candidates. It was not, however, until 1861 that the full honours degree course in the Natural Sciences Tripos came into being, and science could take its rightful place in the formal studies of the University; and it was particularly fitting that Henslow should have been able, in the last few months of his life, to examine in botany the candidates for Honours in the new Tripos, for he had consistently advocated such recognition for science as an important part of modern education at both school and university.

The Act authorising purchase of the new site was passed in 1831, yet the official opening of the 'New Botanic Garden' on this site was delayed for fifteen years. Jenyns refers to 'unavoidable circumstances' causing the delay without throwing any further light. The reason for the delay becomes clear from a reading of the Act; the ground for the new Garden was in fact subject to a 21-year lease entered into by Trinity Hall in 1824, and did not, in the event, become legally available to the University until Michaelmas 1844.

Between 1830, when the University had employed Mr E. Lapidge, a London architect and garden designer, to produce a design for the new Garden, and 1844, when a special Syndicate was appointed, problems of financing the new Garden had apparently been found to be increasingly difficult, and the plan finally carried out developed only the western half of the site. From correspondence between Lapidge and the University in 1840 we learn the reason for the delay. Lapidge is writing to enquire, courteously, whether his bill for professional services was ever going to be paid: 'You will doubtless recollect my drawings for the proposed Botanic Garden at Cambridge, which have lain dormant these ten years, *by reason of the tenant in possession of the land having demanded an extravagant sum for his interest in it.* [My italics] That I should feel very great interest in the work you will rapidly suppose, my plan having been approved by the Syndicate on the 5th October 1830, and my bill of £131.5.0 having remained in abeyance ever since that period. I conceive the tenant's term to be so nearly expired, that the subject could now be advisedly and successfully revived in anticipation of that event.' Lapidge's plea was eventually successful, and a special Grace authorised the payment on 19 February 1841.

The originals of the plans drawn by Lapidge do not seem to have survived, but his manuscript report is in the University Archives, and the general plan to accompany it was fortunately printed in an article in the *Cambridge Portfolio* (Smith, J. J. 1840), with the heading: 'Ground Plan of the Intended Botanical Garden'. Lapidge's scheme envisaged using the whole site, though the easternmost strip (which included, apparently, two smaller properties not included in the 38 acres purchased from Trinity Hall) was to be developed gradually over a longer period as the main Arboretum. A lake with a large island, and an adjacent area of Systematic Beds, were planned in the western part along the Trumpington Road border, and the central feature was to be a formal lawn with glasshouses on the north side. A long containing wall round the east and south sides of the central lawn was intended to demarcate clearly a formal, geometrical inner garden from the more natural garden containing the lake and the Arboretum. How far Henslow influenced the detail of this plan we cannot say, but Lapidge states that he was consulted, and all the main elements in the scheme, especially the Arboretum, are obviously there by Henslow's advocacy or at least with his agreement.

Nowhere amongst the archives has yet emerged any information about the 'extravagant sum' which Mr G. Bullen, the tenant farmer, had demanded from the University to terminate his lease, but we must conclude that the difficulty was wholly unexpected, for a legal 'agreement' exists between Bullen and the University, dated

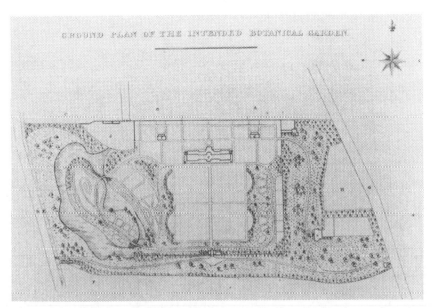

GROUND PLAN OF THE INTENDED BOTANICAL GARDEN.

Figure 48   Lapidge's design for the 'New Botanic Garden' dated 1830. Murray's plan, constructed in 1846–50, contained the main features but compressed into the western half of the site.

November 1830, which was apparently never signed and sealed. The indefinite delay must have been very galling to Henslow, who saw the chance of ever using the 'New Botanic Garden' for teaching now relegated to the distant future. Indeed when, in 1844, the special Syndicate was eventually appointed 'to consider whether and what steps should be taken towards changing the site of the Botanic Garden', Henslow was not initially proposed for membership, and his name was only added in May of that year. Babington, Henslow's pupil and successor, was an original member, and it is clear that he was increasingly deputising for his absent Professor in matters of Cambridge botany. Both must have been responsible for the appointment of the new Curator of the Garden, Andrew Murray, in 1845, who was selected from a field of four candidates after an examination set by Henslow. In addition to a series of written questions, each candidate was asked to produce a plan for the layout of the new Garden, and it seems clear that Murray's plan was not only good enough to win him the post, but also the one he was asked to carry out. By a fortunate stroke of luck, a copy of Murray's plan, long thought to be lost, has recently come to light in the Garden; by comparing this plan with Lapidge's, and with the layout we now have, we can appreciate the changes to the original scheme and also the alterations and additions made by Murray's successors.

The most important change between the Lapidge plan and that eventually carried out by Murray affected the size of the Garden. Murray must have been informed that for financial reasons only the western half of the 38-acre plot could be developed, so that the generous Arboretum planned for the eastern edge of the land disappears, and is only represented by the belt of trees on the east and south sides. It is clear that the other main features of the Lapidge plan – the Lake and the Systematic Beds – which Murray had retained, were in fact carried out without more than modification of detail. Murray, however, had planned a long glasshouse range to run north–south terminating the vista from the main entrance in Trumpington Road – that is in the middle of the site and the Garden as it now is – but this scheme came to nothing with Murray's early death in 1850. The importance of Murray's work for the New Garden was well expressed by Babington, who wrote, in a short tribute published in a local paper: 'All who have seen what he has done in converting a cornfield into a botanical garden containing one of the best collections of hardy trees, shrubs and herbs in the kingdom . . . can and must appreciate his eminence in his professional society.'

Although Henslow remained Professor of Botany for another eleven years until his death in 1861, the story of the development of the 'New Botanic Garden' belongs more naturally to the time of his successor Babington, and will be covered in the next chapter. One event, however, might be selected from the Henslow story, since it so effectively bridges the two worlds in which he operated. On Thursday, 27 July 1854 there was a 'Village Excursion' from Hitcham to Cambridge, led by the Reverend Professor himself, in which no fewer than 287 of his parishioners took part. For this day excursion – made possible by the opening of the new railway station at Cambridge in 1850 – Henslow had prepared a printed booklet of eleven pages, with a large folding plate of illustrations of tropical economic plants to be seen in the Botanic Garden, and a precise time-table for the whole day sightseeing in Cambridge. The party is due to arrive at Cambridge Station at 9.20 a.m.: 'in about five minutes' walk from the . . . Station, we arrive at a back entrance to the New Botanic Garden. About 20 acres are laid out and planted. We pass the spot where the greenhouses and stoves are being erected, to contain plants from hot countries'. After visiting several Colleges and University buildings, they pass rapidly through the Old Botanic Garden and the Anatomical Museum, to be regaled at 2 p.m. to dinner in Downing: 'Assembling in the Hall, we shall find the Vice-Chancellor and his Lady have kindly added to our ordinary frugal fare by a present of a Barrel of Beer, and Plum-puddings enough for the whole party.'

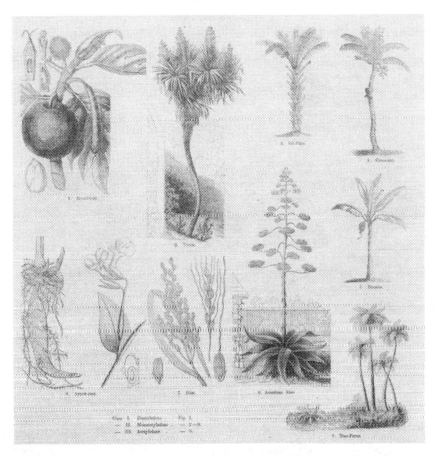

Figure 49   Plate from Henslow's booklet prepared for the day excursion from Hitcham to Cambridge on 27 July 1854. These Tropical Economic Plants were to be seen in the glasshouses of the Old Botanic Garden.

It is difficult to do justice in this small space to the many contributions Henslow made, and our account has barely mentioned his role in establishing natural science, and particularly observational natural history, as an important part of national educational curricula, since these activities were almost wholly conducted outside Cambridge. Perhaps his scrupulous fairness and objectivity, combined with a very real concern to apply his Christian belief to the service of people in all levels of society, were his special, abiding traits of character. Nowhere were these seen to better advantage than when in 1860, less than a year before his death, he found himself in the Chair for the crucial public meeting of the British Association at Oxford, when the famous exchanges took place between Huxley and Wilberforce on Darwinism. Jenyns writes: 'A

Figure 50   Portrait of Henslow by Maguire, 1849.

large audience was drawn together to hear it; and those who were
present on the occasion speak of the admirable tact and judgment
with which he regulated the discussion, showing complete impar-
tiality, allowing everyone fairly to state his opinions, but checking
all irrelevant remarks, and trying to keep down as much as possible
any acrimonious feelings that appeared to mix themselves up with
the arguments of the contending parties.'

We cannot do better than take the last word on Henslow from
Darwin himself. In 1873, in his *Autobiography*, Darwin again recalls
his debt to his old teacher and friend, and assesses his ability: 'His
strongest taste was to draw conclusions from long-continued
minute observations. His judgment was excellent and his whole
mind well-balanced, but I do not suppose that anyone would say
that he possessed much original genius.' This seems, a century
later, a remarkably true assessment, to which we need only add
that, for the progress of science as for the welfare of mankind, a flair
for teaching which releases genius in others may be as important as
the genius itself. Without Henslows there are no Darwins.

6

Babington, Vines and Lynch: the fragmentation of botany

Figure 51 Bridge of Sighs, St John's College, with *Elodea canadensis*. Babington was accused of introducing this American water-weed into the River Cam.

Babington and the Herbarium

The first half of the nineteenth century, in Cambridge as in Britain generally, saw the rise of the professional natural science which is familiar to us in universities, museums and scientific institutions at the present day. In this great movement, as we have seen, John Stevens Henslow had played a modest but important part, firstly as an enthusiastic teacher in Cambridge, and in his later years as an adviser on botany, and science in general, in national school and university curricula. Allen (1976) has described the 'Cambridge Network' of scientists, historians and others who helped to bring about 'the professionalisation of English science', and has pointed out that from this loosely-knit pressure group might be traced, flowering in the second half of the century, 'those closely-meshed dynasties of scholarly minded kinsfolk that Noel Annan has termed the "Intellectual Aristocracy".' He distinguishes three of these related family groups, all with characteristic, important contributions to science in general and biology in particular. The first is composed of leading Quaker families, such as Buxton, Barclay and Cadbury, with a continuous tradition of natural history. The second is 'formed by the families of devout Evangelicals descended from the original Clapham Sect', one group of which contains the linked family names of Macaulay, Trevelyan and Babington. The third dynasty, by far the most famous, radiates from the Hookers, father and son, who built the international reputation of Kew. The younger Hooker's first wife was Henslow's daughter and, as we have seen, Henslow's wife was sister of Leonard Jenyns, who married into the family of Daubeny, the Professor of Botany in Oxford.

These groups of talented inter-related families seem to have been uniquely characteristic of the Victorian era. Discussing their significance for natural history studies, Allen says: 'This degree of interlocking conferred . . . certain advantages. For a start, it ensured a continual traffic in news; one field knew for much of the

time what was occurring in another and, in the process, often drew
on its ideas or copied features of its institutions. It brought, too, an
underlying stability in morale. The consciousness that he worked
within an accepted family tradition freed a naturalist from social
isolation and rendered him immune to accusations of eccentricity.
It even had administrative convenience. The passing of scarce jobs
from fathers to sons . . . at least guaranteed a singular continuity of
outlook and, in an age when poor men in ill-paid posts were so
frequently undependable, meant that certain standards of probity
and rectitude could be relied on to be observed simply out of regard
for family tradition.' It was, of course, in this 'golden age' of
personal contact and efficient, if verbose, communication by letter
that Charles Darwin's genius flourished.

Against this background of a gifted intellectual élite we must
now look at the broader canvas of European scientific thought, at
least as it affects the development of botany. The Linnaean 'revolution' had come and gone, leaving behind a robust, efficient, international Latin terminology and nomenclature which survives to the
present day. From the burst of energy shown in Cambridge by the
young Henslow, a new practical science was taking shape, in which
mere learning by rote of names of genera and their floral characters
was rejected as wholly inadequate, and the student was encouraged
to find out for himself by formal practical studies on living plants.
Taxonomy (incidentally, a word first used by de Candolle and
quickly adopted by Henslow) was seen in its rightful place as a
necessary 'backbone' to scientific communication, but *not* as the
science itself. The creation of the new Botanic Garden in Cambridge was part of that process, which had its counterpart in universities in many parts of Europe.

In one important respect, however, English universities were
seriously isolated. The effect of the Napoleonic Wars, combined
with the rise of the British Empire, had turned much of the energy
and talent of the nation away from continental Europe, and had in
particular favoured a linguistic laziness which has, indeed, persisted
to the present day. Where John Ray spent more than two years in
continental travel, and Richard Bradley had derived great stimulus
from his long visits to Holland and France, and Thomas Martyn
had written tourist guides to Italy based on personal experience,
neither Henslow nor Babington ventured outside the British Isles
for more than very brief periods, and neither, apparently, could
converse in any modern foreign language except a little French.
French, indeed, was the link by which Henslow appreciated the
new science of continental Europe in the works of de Candolle.
Contact with German scientific literature was only possible for him
when a particular work (such as that of Liebig on soil chemistry)
was already available in translation into English or French.

This ignorance of the German world of learning broke down quite suddenly about the time that Henslow ceased to have much influence in the University, a rapid change which owed much to the benign influence of Albert, Prince Consort, patron of science in general and natural history (biology and geology) in particular. The fruits of this Germanic patronage and enlightenment were mainly at first concentrated in London, where, for example, the great work of Hofmeister, first translated into English in 1862, helped to lay the foundation for modern comparative anatomy with its evolutionary interpretation of the linear series of organisms in the plant kingdom. By 1872, Thomas Henry Huxley was teaching the new 'laboratory botany' at South Kensington to an enthusiastic group of students with the assistance of William Thiselton-Dyer and Sydney Vines, and it was Thiselton-Dyer who took over the course when Huxley was ill in the following year. As Gilmour (1944) puts it: 'It is not easy for us to recapture the intense excitement of these classes. Huxley's team faced the difficulties of all pioneers. Botanical teaching at that time was practically confined to the systematics of flowering plants, and preserved material for demonstrating the morphology and anatomy of seaweeds, fungi and other "lower" plants was non-existent.' 'I was generally up half the night', says Dyer, 'rehearsing the demonstrations for the next day.' Marshall Ward, one of the keenest students, who (see chapter 7) became Professor of Botany in Cambridge after Babington, fainted at his work from over-excitement!

During these exciting years of the Darwinian revolution and the establishment of laboratory science, what was happening at Cambridge? It is sad to have to record that the great promise of Henslow's 'new botany' had come to nothing. By the time Babington succeeded to the Chair, it seems that whatever enthusiasm for new ideas he may have had as a keen pupil of Henslow over thirty years before was now replaced by a suspicion and mistrust of the young men who surrounded him and troubled his old age with new methods and new interests. But to understand his resistance, if not to approve it, we should look more closely at the circumstances of his family background and Cambridge career.

Charles Cardale Babington was born in Ludlow, Shropshire in 1808. His father, Joseph, and three of his father's brothers had all been educated at St John's College in the 1770s and 1780s, and the family seat was at Rothley in Leicestershire. His childhood was spent in several counties – Leicestershire, Nottinghamshire, Wiltshire – and finally the family moved to Bath in 1822 after Joseph had resigned from the church because of ill-health. In his Journal Babington records that when he was ten years old his father taught him 'the elements of botany from Lees' "Introduction" and Withering's "Arrangement".' Part of his education was at Charterhouse

but, according to his own assessment, he learned most at a private
school in Bath from 1822 to 1825, where he was equipped for his
Cambridge career. Of those days he wrote: 'during the years I was
at that school, as a day scholar, I formed an intimate acquaintance
with the neighbourhood of Bath and began to study its botany and
to collect plants and insects.' This schoolboy botanising, indeed,
gave rise to his first published book, *Flora Bathoniensis*, which
appeared in 1834.

So far, the young Babington's education had been very similar to
that of Henslow, only twelve years older than himself. Both came
from good professional homes where learning was appreciated and
where father enjoyed and encouraged his son's interest in natural
history. Perhaps the only difference was that Babington was a keen
field botanist (as well as an entomologist) while still at school,
whereas, as we have seen, Henslow learned his botany late. In
Cambridge, however, things were very different for the two men.
Henslow had found a vacuum which he saw as a challenge, and
splendidly filled it, converting keen young men both to the
pleasures of field natural history and the new scientific outlook.
Babington was one of his first 'converts', recording in his Journal
that he attended Henslow's 'first lecture on Botany' on 30 April
1827; then, two days later, we read the following: 'Conversed with
him after the botanical lecture, and was asked to his house . . .
Assisted Professor Henslow in putting his things in order, before
and after his lectures.' In fact, he sat at Henslow's feet for six
successive series of Easter Term Lectures, and took part in most of
the local excursions, recording, for example, his participation in the
Gamlingay excursion on 19 May 1831, followed by 'Henslow's
party to Wood Ditton' three days later. Even if Babington had been
ambitious – and there is little to suggest that he ever was – Henslow
was occupying the only Chair which he would have wanted.

It is not, therefore, surprising that, after taking his M.A. in 1833,
Babington settled down to a comfortable bachelor existence in his
rooms in St John's College, writing his books, enjoying field
studies in natural history and, later, local archaeology, and gradually becoming Henslow's assistant and deputy, in fact, if not in
name. Three local Societies, all of which are still in existence, owe
their origin, in large part, to Babington's stimulus and support: the
Ray Club, founded in 1837 'for the cultivation of natural science by
means of friendly intercourse and mutual instruction'; the Entomological Society, later to become the Cambridge Natural History
Society; and the Cambridge Antiquarian Society. As Liveing, Professor of Chemistry, and a fellow Johnian, wrote in an obituary in
the *Cambridge Review* in October 1895: 'the condition of the University at that time was so utterly unlike what it is now, that the

younger men amongst us will perhaps find a difficulty in crediting what I have to tell of it . . . There was absolutely no opening for those who followed after Natural Science . . . At the time of which I am writing, Henslow had gone down to a living in Suffolk, and it was Babington more than anyone else who drew around him the young men, and the older ones too, who took pleasure in Natural History . . . He knew all the haunts of plants and insects in the county, and it was a pleasure to him and to the four or five who sometimes accompanied him on a long day's ramble, to try to find something new to him as a denizen of the locality we were exploring.'

The picture we have, then, of Babington in his middle age is that of a kindly, learned bachelor don (though not yet a Fellow of his College) studying local history, archaeology and biology, and travelling widely in the British Isles in the vacations to collect material for his first important book, the *Manual of British Botany*. This is a concise, scholarly and up-to-date new British Flora; it appeared in 1843 and ran to ten editions, the last by A. J. Wilmott in 1922. The *Manual* established his reputation nationally and even internationally, because it made a conscious attempt to take into account the most important taxonomic information which had been published in Germany by Koch, Reichenbach, Sturm and others. The isolation from the continent, so serious as we have seen for the new experimental science, was easier to correct in systematic botany because of the common Latin language, if not for all the text, at least for the binomial nomenclature and many of the technical terms. As now, all that was necessary to read German systematic works was a limited knowledge of the written language, which Babington apparently acquired without much difficulty.

Figure 52   Yew Tree (*Taxus baccata* var. *dovastoniana*) planted by Babington in 1843 on the Backs of St John's College. Photographed July 1980.

In 1851 Babington was elected a Fellow of the Royal Society. By that time he was in the forefront of botany nationally, and was playing a prominent, though always somewhat unobtrusive, part in bodies such as the British Association, the Ray Society and the Linnean Society. His contributions to learned journals were very numerous indeed – 132 papers are listed in the Bibliography to his *Memoirs* (1897), and there are in addition 52 'communications' in the *Proceedings of the Cambridge Antiquarian Society*. Of the later published books, the best-known is his *Flora of Cambridgeshire* (1860), which conveniently brings together all the previous records, together with the results of his own and Henslow's active botanising over the previous thirty-five years. This book, the fifth in the long line of Cambridgeshire Floras, was published exactly 200 years after Ray's pioneer 'Cambridge Catalogue'; it was to remain an accurate guide to Cambridgeshire plants for another century, and did not significantly 'date' because the great changes in agricultural practice which resulted, for example, in the ploughing of most of the chalklands had happened early in the nineteenth century, and were already history when Babington wrote.

Would it have made much difference to the development of botany in Cambridge if Henslow had resigned in, say, 1846 when the New Garden was opened, and allowed Babington to take the Chair when he was still, by academic standards, a young man? We can certainly feel sorry for Babington that this did not happen; but it seems doubtful whether he would ever have shown the wide sympathy for the 'new science' which Henslow had, and events might not have been very different. In particular, the postponement of radical changes in University and College organisation would have continued to act as a brake on the development of an experimental natural science, whatever the Professor of Botany might or might not prefer to teach. With respect to the Garden, however, Henslow's departure was most unfortunate. We have seen that Babington played an important part in the planning and layout of the New Garden from 1845, and although there is little evidence that he would ever have shared Henslow's wider vision of the range of purposes the Garden should serve, he would surely, as Professor of Botany, have used the Garden more both for teaching and for his own research into the taxonomy of the British and European floras if he had been encouraged by circumstances to do so in the early formative years from 1845. As it was, he put more and more of his energy into collecting specimens for the University Herbarium, and gradually became a herbarium taxonomist to whom the rest of botany, even the Garden, was at best mildly interesting and at worst an alien and disruptive growth.

Vines and experimental botany

So began the melancholy story of the long fight to establish in Cambridge a modern experimental botanical science, a fight which succeeded almost literally 'over Babington's dead body', with the appointment to the vacant Chair in 1895 of Harry Marshall Ward, the enthusiastic pupil of Huxley at South Kensington. A new century brought a new era and a new Botany School now wholly independent, both physically and intellectually, of the Botanic Garden (see chapter 7). In this long struggle Sydney Vines had taken a leading part, establishing gradually against considerable opposition a place for experimental botany in Cambridge. Although the details of this story must lie outside the scope of my book, some short account of Vines' career in Cambridge is necessary: for more detail, see Green (1914).

Vines entered Christ's College in 1872, after several years in

London studying medicine. He was attracted to Cambridge by Michael Foster, another pupil of Huxley, who had already formed at Trinity a school of experimental biology, and was lecturing on animal physiology to audiences of enthusiastic young biologists. Vines' interest in botany arose from his association with Thiselton-Dyer, with whom he ran the South Kensington botany course in 1875, and his increasing concentration on plant physiology in Cambridge developed naturally as a complement to Foster's lectures. Like Foster, Vines was severely handicapped by the absence of laboratory facilities, and he had to deliver his first lecture course in Cambridge in 1876 without any accompanying practical classes. In spite of this initial handicap, he attracted keen students, amongst them Bower, who became the famous morphologist and Professor of Botany in Glasgow, and has left us, in his autobiographical *Sixty Years of Botany in Britain*, published in 1938, a very readable account of the 'impressions of an eye-witness' who entered Trinity College, Cambridge, in 1874 with the intention of specialising in botany. By the time Vines had achieved some University as well as College recognition in 1880 he had built an influential school of plant physiology which has flourished in Cambridge to the present day. This school had learned directly from German scholarship, and Vines acted as the pioneer in this respect, studying under Sachs in Würzburg in the Long Vacations. In 1880 he published a translation of Prantl's *Elementary Botany* which provided, as Vines put it in the Preface, 'A work on botany which, while less voluminous than the well-known *Lehrbuch* of Professor Sachs, should resemble it in its mode of treatment of the subject, and should serve as an introduction to it'.

To appreciate the revolution in teaching represented by the Prantl–Vines text-book, we can briefly classify its contents, which are clearly set out as follows: Part I Morphology, 23 pages; Part II Anatomy, 44 pages; Part III Physiology, including Reproduction, 29 pages; and Part IV Classification, 182 pages, 60 of which are devoted to the Lower Plants. From being almost the whole of botany, as it still was to Babington, the systematics of Angiosperms has dropped to about one-third of the course. Until very recent years, this is roughly the position it has occupied throughout the present century in university courses of botany in England. If our comparison, however, is made, not with the contents of Babington's lectures at the time of Vines, but with Henslow's own elementary text-book nearly fifty years *earlier*, we find the 'revolution' by no means so obvious. In Henslow's book (1836), Part I, 'Descriptive Botany' including morphology and systematics, occupies the first 150 pages, but the whole of Part II, of very similar length, is devoted to 'Physiological Botany': in other words,

Henslow had foreshadowed the dethronement of Linnaean systematics, but circumstances prevented him from establishing the 'new science' as a viable tradition in Cambridge. In this respect, the version of the struggle between Babington and Vines given by Green (1914) is unfair to Henslow. Where Green says (p. 539) that 'Babington was old and feeble, and his personal interests were those of a specialist in particular groups' we can have no quarrel with him. But in the next sentence: 'Outside Taxonomy he did not care to go, and in his teaching he was content to plod along the old lines laid down by Henslow', he is surely unfair. Everything suggests that Henslow would have welcomed the arrival of the new experimental science in Cambridge, and given what help he could in establishing it.

The later career of Vines as Professor of Botany in Oxford, and the role of Francis Darwin, son of Charles, who became Deputy Professor four years before Babington died, in building the school of plant physiology at Cambridge, must lie outside our story. We can, however, note that the painful birth of experimental biology in Cambridge produced a fierce rejection of the old botanical traditions which more recent generations have had to try to live with. We shall look at some implications of this in the succeeding chapters.

The 'New Botanic Garden' and the career of Lynch

We have seen in the previous chapter how the New Garden was planned, postponed and eventually laid out, and how, although Henslow's keen interest had been reduced by 1846 to a sympathetic and rather distant concern, the eventual design incorporated all the elements which he had advocated, based especially on Kew and the advice freely given to him by the Hookers. The first Curator of the New Garden, Andrew Murray, responsible for the detailed design, was obviously a man of considerable knowledge and practical skill, and his early death was a very unfortunate loss.

The establishment of the New Garden took place in three phases: the purchase of the land in 1831, which we commemorate with this book; the establishment of the Garden's main features, which included the planting of the Arboretum and the transfer of hardy herbaceous plants from the Old Garden to the new Systematic Beds; and finally the building of the glasshouse range and the transfer of the remaining stock from the old greenhouses to the new. These three processes spanned a little over a quarter of a century, and it was not until the last transfers were finished, under the Curatorship of Murray's successor, James Stratton, that the University appointed a new Syndicate 'for the future management of the Garden'. This body has functioned ever since, with some alterations to its regulations, and there is an unbroken series of Annual Reports from the first one in 1856 to the present day.

In their first Annual Report, dated 5 March 1856, the Syndicate 'think it desirable to give a more detailed account of the character and state of the Garden than will be necessary in future years', and make the following admirably clear statement of the purpose of the Garden: 'In the formation of the new Garden, it was intended to make science the first consideration; but in order to encourage a general taste for botanical studies, and to render the Garden an agreeable acquisition to the University, the designer consulted ornamental appearance wherever it did not interfere with the main object.' They then proceed to make a virtue of necessity: 'the funds applicable to the support of the Garden being small, its objects are necessarily limited . . . The formation of a collection of hardy plants was thus preferred to one consisting of numerous tender species, which would have required the erection of extensive and costly plant-houses for their preservation.' Already the trees, planted according to the Henslow scheme from 1845 on, are seen as one of the potentially great features of the Garden: 'The trees form a belt surrounding the whole of the ground, to which they will ultimately be a considerable protection. They are arranged as far as possible on a scientific plan . . . Amongst them will be found nearly all the trees that will stand our climate, and it is believed that, when grown up, they will form one of the most perfect Arboreta in the kingdom.' In contrast to the tree planting, which had to be an imaginative act of faith, the Systematic Beds were quickly and effectively established, and indeed have changed remarkably little since their inception. Here the Report says: 'The Herbaceous Plants are arranged in beds on the right of the great walk. Each bed contains a single natural order, or a division of one. This is a very perfect and valuable collection, and it is much used by the members of the University who study scientific Botany.'

This first Report covers, in effect, the establishment of most of the main features of the nineteenth-century Garden as we see it today, a period entirely under the Curatorships of Murray and Stratton. Stratton added the present Lake which was constructed in 1858–9 at a cost of £190. In 1864, on the death of Stratton, the Syndicate appointed as Curator William Mudd, a man of unusual talent whose horticultural training was in private gardens in the north of England. Self-educated in botany relatively late in life, largely through contact with enthusiastic naturalists on the teaching staff of the Quaker School at Great Ayton, he rapidly became an authority on the local flora, and acquired a keen and permanent interest in lichens which culminated in the publication in Darlington in 1861 of his classical work *Manual of British Lichens*. As his *Gardeners' Chronicle* obituarist puts it: 'If anyone will look through this, remembering that it is the production of a man who had to educate himself after reaching mature life, and who at the time that he was engaged upon it was working hard with his hands for twelve hours a day, and keeping a wife and family upon a wage of something like 25s a week, he will see that the book is really a wonderful monument of energy and perseverance.'

Unfortunately this early promise seemed to come to nothing in Cambridge. The difficulty may have been partly caused by his state of health, which had apparently been seriously affected by his overwork at microscopy of lichens before he took the Cambridge post; it seems likely, however, that he found the atmosphere of the University and his social position in town and University so alien to his experience that he could make little of it. It seems particularly sad that he did little or nothing to keep up his interest in field botany, where his earlier enthusiasm for lichens might have made a really valuable contribution. He seems to have increasingly exercised his botanical talents and, no doubt, usefully supplemented his meagre salary, by taking in private pupils for coaching, a practice which was not welcomed by authority. Indeed, after his death in 1879, we read that the Syndicate 'after careful consideration, have come to the conclusion that the Curator ought to devote his whole time and attention to the Garden, and that it is not desirable that he should take private pupils'. The Victorian explorer Alfred Maudslay records how in 1868 he went to Cambridge to take the Natural Science Tripos, where he was coached in Botany by Mudd, 'an illiterate Scotchman (*sic*) who smoked very strong tobacco and smelt strongly of whisky'. Apparently Mudd thought little of Darwinism, and gave it as his opinion that 'that man Darwin will go to Hell'. (Boutilier, 1975).

Figure 53   Staff of the Botanic Garden in 1876. The Curator, William Mudd, is wearing the top-hat.

One of the few developments recorded in Syndicate Reports during Mudd's Curatorship was the establishment of the allotments, let out on annual tenancies, on what was up to 1867 referred to as 'the field adjacent to the Botanic Garden'. This use of the undeveloped eastern part of the Garden continued unbroken until 1950, from which date the 'new area' was finally developed using the Cory Bequest. Towards the end of Mudd's career, the glasshouse range designed by Stratton was proving more and more troublesome to heat and maintain, and in 1877 we read in the Annual Report that 'the gale during the winter would have blown down the palm-house if it had not been supported by an adjoining building: it therefore urgently needs repair'. The impression conveyed by this terse Report is that the problem was not entirely a financial one and that if the Syndicate had been honest they would have levelled some criticism at the Curator. Nevertheless, all was not deterioration and decay. In 1877–8 two new propagating houses were built, partially replacing old houses which had themselves been constructed out of materials from the glasshouses at the Old Garden some twenty-five years earlier, and we read in that year's report (1878) that 'more than 150 members of the University have used the Garden for the study of Botany, and many thousands of specimens have been furnished to teachers of Botany'.

By the time Lynch was appointed in 1879, the Garden was set to function in the way which has continued to the present day. It exhibited a teaching collection of living plants with special emphasis on the hardy species, and it had begun to supply an increasing amount of material for both practical classes (already established by Vines) and examinations. The Lynch appointment practically coincided with the new scientific teaching for the Natural Science Tripos, and represented an opportunity for building up the original Henslow conception in the new climate of opinion. We can now see to what extent Lynch was able to do this.

Richard Irwin Lynch was born in Cornwall in 1850, son of a head gardener on a private estate, and trained as a gardener at Kew, where he became foreman of the Herbaceous Department at the age of 21 years. He was only 29 years of age when he took up his appointment in Cambridge, and seems to have wasted no time in tackling some of the more obvious areas of neglect, such as the condition of the glasshouse collections. His arrival in Cambridge coincided with an unprecedented flooding of the low-lying part of the Garden after a very heavy thunderstorm on the night of 2–3 August, and his first winter in Cambridge was unnaturally severe, with hard frost before the end of November, and a reading below 0°F near Cambridge on the night of 6–7 December.

The Report of the Syndicate for 1881 records the completion of the Curator's House, the building of which had been decided on when the opportunity to provide a resident Curator had arisen with Mudd's death two years previously. Lynch occupied the house continuously until his retirement in 1919, and since that time it has continued to be the residence of the Superintendent of the Garden. By the autumn of 1881 Lynch had weighed up the needs of the Garden sufficiently well to approach the Vice-Chancellor, as Chairman of the Syndicate, with a definite proposal that a special grant be made available for each of three successive years in order to employ 'from two to four additional men' to work on the main woody collections. This letter, together with a short covering note by Babington as Professor of Botany, is published as a special report to the Senate on 13 March 1882. It is an admirably clear request in which, without blaming his predecessors except by implication,

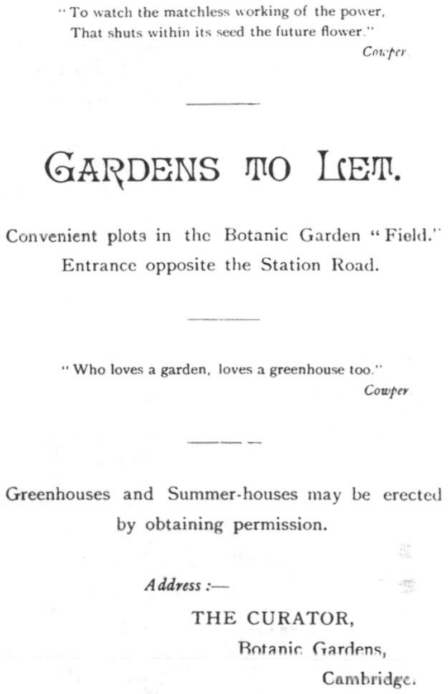

"To watch the matchless working of the power,
That shuts within its seed the future flower."
Cowper

GARDENS TO LET.

Convenient plots in the Botanic Garden "Field."
Entrance opposite the Station Road.

"Who loves a garden, loves a greenhouse too."
Cowper

Greenhouses and Summer-houses may be erected by obtaining permission.

Address :—

THE CURATOR,
Botanic Gardens,
Cambridge.

Figure 54   Allotment advertisement from the time of Lynch.

Lynch explains that the woody collections have been neglected and that Henslow's great design is in jeopardy. He says further: 'the purchase of a small number of trees – not obtainable by exchange – would be desirable, as the collection is already a fine one and might be made a speciality, being one of the cheapest to maintain. This collection is allowed by all authorities to be an important feature in a Botanic Garden. We have rare species and fine specimens for which our collection has already some reputation.' Babington's contribution to the exercise seems to have been to persuade Lynch not to ask for 'a much larger sum' because of the present state of the University funds – an act of practical diplomacy which, as any modern Head of Department knows, could well have been very sound advice. Anyway, Lynch got his money – £300 a year for three years – to employ extra gardeners, who carried out the necessary arboriculture, and tackled the arrears in general maintenance. From then on, throughout the whole of his career, Lynch developed the woody collection, and obviously grew to love the individual trees, eventually publishing (in 1915) his account of the *Trees of the Cambridge Botanic Garden* which summarises the information he had collected together with such careful enthusiasm.

A question which came up at this time, exactly one hundred years ago, was that of summer Sunday opening. After a strongly-conducted campaign on both sides, the Grace allowing Members of the Senate 'with their friends' to visit the Garden on summer Sunday afternoons was adopted by 144 votes to 129. Babington, a strict Evangelical Sabbatarian, gave notice in a circulated fly-sheet that he intended to vote against Sunday opening on conscientious grounds, and his views make interesting reading today. It is particularly worth noting that the question being discussed was *not* whether the Garden should be open to the public on summer Sundays – a change which in fact did not come about until 1975, and is still 'experimental' and subject to annual review – but whether senior members of the University might be allowed in on that day. No one apparently raised the question of what benefit the general public might derive from such a privilege. In the event, the summer Sunday opening was dropped after three seasons, and did not start again until 1900, when something like the present Sunday key system was introduced.

After giving first priority to the existing collections, Lynch then began to plan and carry out innovations. In 1882 he re-designed the 'bog and water garden' area, and laid out a new bed for medicinal plants; in 1883 he reconstructed the small, unsatisfactory rock garden (now part of the Terrace Garden), and in the following year established what he claimed to be the 'first ornamental Bamboo collection' in the position where we still have it. Soon after this, about 1886, he began to establish his collection of hardy cacti, which later became one of his special interests and from which we derive that most famous of all Cambridge Botanic Garden plants which he named *Opuntia cantabrigiensis* in 1903. From 1882 onwards he began to build up a collection of ferns, with the expressed intention of making at least the British material as complete as possible; how far he succeeded can be judged from his *List of Ferns and Fern Allies cultivated in the University Botanic Gardens* (sic), *Cambridge*, a rare pamphlet published in 1897, in which are listed 391 species, as well as many varieties.

Whether Lynch expected to receive much guidance from either the Syndicate or the Professor of Botany in determining his priorities, it is very difficult to say, but he must quite quickly have decided that it was up to him so far as broad policy as well as detailed decisions were concerned. Babington must already by 1879 have been little use to him, except for advice on obtaining extra money for his initial 'rescue' schemes, and Lynch was expressly forbidden, after the unfortunate experience with his predecessor, from doing any private coaching. In the circumstances, he turned more and more to scientific horticulture, and developed friendships with other enthusiastic gardeners, both amateur and professional.

Figure 55 Lynch's *Opuntia cantabrigiensis*, still grown in the Garden.

One of the most interesting of these acquaintances was the physiologist Michael Foster, later Professor Sir Michael Foster, F.R.S., whom we have already mentioned as a colleague of Vines, in his capacity as enthusiastic amateur gardener and Fellow of the Royal Horticultural Society. Both Foster and Vines were members of the Syndicate in the mid-eighties, and Foster is listed annually for several years from 1881 as having presented plants to the Garden. Lynch must, in particular, have derived his enthusiasm for the genus *Iris* from Foster, to whom he dedicated the *Book of the Iris*, his only substantial published work, which appeared in 1904.

By 1885 it had become apparent to Lynch that his next task was to persuade the University to replace the ageing and decrepit glasshouse range, which Stratton had constructed, with a new range in which great improvements in both layout and heating might be incorporated. By June 1887 he had so effectively pleaded his case that a Grace was passed by the Senate in that month authorising the Syndicate 'to obtain a detailed plan . . . for new houses and a pit' and also 'for a temporary Research Laboratory . . . in the Botanic Gardens', the cost of the glasshouse range not to exceed £3000, and that of the laboratory £250. A specially appointed Sub-syndicate recommended to the Syndicate the acceptance of plans prepared by the firm of Boyd, of Paisley, having held consultations with Mr James Boyd; 'at their request Mr Lynch visited the Paisley works and spent three days in going into the practical details'. Thus the

Figure 56   Staff of the Botanic Garden in 1912. The Curator, Richard Irwin Lynch, is holding his hat.

Figure 57   *Thladiantha dubia*, still growing in the Botanic Garden, originally sent by Vines to Lynch as an 'experimental plant' in 1880.

present glasshouse range was planned, and constructed between 1888 and 1892. Prophetically the Annual Report for 1888 states that 'the Syndicate have reason to believe that the material used and the methods of construction are thoroughly sound and satisfactory. Solid foundations have been laid so that years hence, when the woodwork must be replaced, it may be rebuilt on the same walls.' This is in fact what happened when the range was re-built in teak in 1934–5.

Judging from the state of the Garden archives today, Lynch's appointment brought order where there was serious neglect in all parts of the Garden. Thus the surviving Garden correspondence begins abruptly in 1879, and the system of Entry Books, in which numbers are assigned to acquisitions of plants and seeds and the details recorded, dates also from this time and is in fact continuous to the present day. A comparison of the seed and plant exchanges recorded in the Annual Reports of 1876 and 1886 reveal that Lynch had reactivated a moribund tradition to a remarkable degree. In 1876, under Mudd, only three donations of plants and seeds are recorded, and there is no mention of any seed exchange with other Botanic Gardens, but in contrast in 1886 the Annual Report records 1319 plants and 1047 seed packets received and lists 24 Botanic Gardens, 56 individuals and 7 nurserymen as the donors. The same Report mentions that the first printed Seed List was produced by the Garden in this year, and was sent out to 87 correspondents, 34 of whom requested, and were sent, 1358 packets of seeds. As with the Entry Books, so with the Seed List, the system established by Lynch has operated continuously up to the present day, and remains an important link with Botanic Gardens throughout the world.

During the 1880s Lynch's reputation grew nationally and even internationally as a plantsman-horticulturist who combined practical gardening skills with a very extensive and indeed fast-growing knowledge about the world's flora as he could learn it from plants in cultivation. In 1884 he attended the International Congress of Botany and Horticulture at St Petersburg (now Leningrad), where he gave two short papers, one on growing wetland plants in a small water garden, and the other on the root-tubers of *Thladiantha dubia*. (It is good to find that we still have both Lynch's Water Garden and Lynch's *Thladiantha* today.) He obviously used the Congress not only to strengthen the exchange of material between Cambridge and the famous St Petersburg Botanic Garden, but also to make personal friends with botanists and horticulturists throughout Europe. Two letters to Lynch from the famous Russian botanist Maximovicz dated 17 September 1884 and 20 June 1885, preserved in the Garden archives, attest to this fact; Maximovicz writes not

Figure 58   The glasshouse range in 1911. In the foreground is the old rock garden, now the Terrace Garden.

only of business matters such as reprints of Lynch's Congress papers, but also says: 'my daughter and I are truly obliged for your kind remembrance and the fulfilment of a promise made, I thought, more in jest than in earnest. She delights in the idea to try these receipts [presumably cookery recipes] in succession, beginning with the easiest.' Many of us can recall such friendly promises made to the wives and daughters of colleagues at international meetings, but we are not all as efficient as Lynch at honouring them!

As a plantsman, Lynch seems to have been unusually eclectic. Although he developed quite early some special favourites among horticultural genera – *Iris* and *Paeonia* are the best examples – he remained interested in all new and unusual plants which he could acquire for the Cambridge garden, and obviously treated difficult subjects, such as the carnivorous aquatic *Aldrovanda*, of which he obtained material in 1895 and again in 1906, as challenges to be accepted and (usually) overcome. The successive Annual Reports of the Lynch period list many such 'choice subjects', often with short comments explaining their special scientific, economic or popular interest. Thus in the Report for the year 1894 we read that the Garden received '*Centaurea crassifolia*, the only native of Malta peculiar to that island' and in 1913 Lynch supplied the plant in flower for the illustration in the *Botanical Magazine* (t. 8508). This plant we are now growing once more, under the name *Palaeocyanus crassifolius*, as an endangered endemic species. To take another example, the Report for 1897 records the flowering in the Garden of '*Buddleia variabilis*, a handsome new species (figured from Kew in

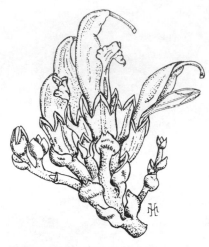

Figure 59 *Lathraea clandestina*, a re-markable early-flowering parasitic plant, first grown by Lynch, and planted by him outside the Garden about 1906. It is now well established by the River Cam.

the *Botanical Magazine*, t. 7609)', which is *B. davidii*, the familiar garden Buddleia so widely grown and naturalised in Britain at the present day.

Some of Lynch's new plants developed under his care and enthusiasm into horticultural success stories which are still with us today. In the Report for the year 1887 he heads a list of plants newly acquired with '*Gerbera Jamesoni*, a fine *Composita* introduced from the Cape'. From this beginning, by a programme of crossing and selection Lynch produced a range of showy, 'improved' cultivars which are the foundation of the modern Gerberas. (There is an article by Lynch, describing his work and illustrated by a fine colour plate, in *Flora and Sylva* for the year 1905.) Both the Royal Horticultural Society and the University of Cambridge formally recognised his outstanding achievements in horticulture in the same year, 1906, the former giving him the Victoria Medal of Honour, and the latter an Honorary M.A. degree.

Before the end of the nineteenth century, the Cambridge Botanic Garden under Lynch achieved a reputation second only to that of Kew, and was providing a growing service to the University with outstanding efficiency. After 1883, when Vines was made University Reader in Botany, and Francis Darwin and Walter Gardiner joined him as University Lecturer and Demonstrator respectively, regular practical classes in the modern manner were increasingly available, and in 1887 a new, though temporary, laboratory with accommodation for about 100 students was built by the side of Babington's Herbarium. Gardiner's *Syllabus of a Course of Practical Botany for use at the Botanical Laboratory, Cambridge*, published in two parts in 1890 and 1891, sets out in detail a course which uses a great deal of living plant material, most of it obviously from the Botanic Garden, and the following is included in the notes to students: 'xi. Students should make a point of visiting the Botanic Gardens (*sic*) in order to become thoroughly acquainted with the plants as *living* plants. They may also obtain specimens of flowers for the purposes of study by applying to the Curator.' Thus began the regular servicing of the practical classes in botany which, again, has continued as an unbroken tradition to the present day; although the balance of the modern courses, and in particular the rise of cell biology, meant that the numbers of specimens supplied by the Garden probably passed through a maximum in the period after the opening of the new Botany School (the present building on the New Downing Site) in 1904 and up to the outbreak of the First World War. Lynch records that in 1908 he supplied 101,471 plant specimens for 'Class Work' (a term still used) and in 1909 the figure was 108,979.

Together with the provision of regular undergraduate teaching

in Botany for the Natural Science Tripos arose the need for research
accommodation, and, as we have seen, a small laboratory was built
behind and connected to the new glasshouse range; this building,
which in the 1940s was used as a lecture room and practical laboratory by Gilbert-Carter when teaching systematic botany in the Part
II Botany Course, has in recent years become the Garden work-
shop. From this small beginning in 1890 developed a considerable
research use of the Garden, especially by Bateson and his associates
in the early days of genetics, and it is clear that, initially at least,
Lynch welcomed this development, recording with pride that, in
the year 1903, 'thirteen investigators have been engaged on work
requiring the cultivation of plants on a large scale'. Moreover,
Lynch was not merely providing facilities for research workers to
grow their own plants; in the case of the investigation of the garden
Cineraria, where the young Bateson took issue with Thiselton-
Dyer in a correspondence which enlivened the pages of *Nature* in
1895, Lynch undertook at Bateson's request an extensive programme of what we would now call 'biosystematic' or 'experimental taxonomic' investigation of *Senecio cruentus* and related
species, and published the results in 1900.

Much of the early genetical research of Bateson and his colleagues
involved growing large populations of the species under investigation, and increasingly in the early years of this century allotment
land was being used, and experimental plots established on land
owned by the Garden. How fortunate was this collaboration between Bateson and Lynch can be judged by the fact that, when a
Grace came before the Senate to authorise the sale of $1\frac{3}{4}$ acres of the
'unused' land in the northeast corner of the Garden for £7,000,
Bateson led the opposition which resulted in the defeat of the
proposal. The fly-sheet published by Bateson, dated 1 March 1904,
is extraordinarily prophetic, one paragraph in particular being of
outstanding interest: 'At the discussion in the Senate House I named
one object to which a section of the estate should be devoted. An
experimental garden for research and for demonstration is becoming an indispensable part of the equipment of a great biological
school. Such a garden is practically a laboratory and must be as
accessible as the laboratories. This institution should be developed
gradually in conjunction with the Botanic Garden, nor can it be
established elsewhere without heavy initial cost and a larger permanent endowment. The work contemplated will pertain not to
Botany alone, but to all the divisions of biological science –
Zoology, Botany, Physiology, Pathology. Though for such a purpose it is obvious that only a part of the 17 acres will be required, no
prophetic gift is needed to foresee academic uses for the remainder.'
Fifty years after Bateson's successful defence of the allotment land,

Figure 60   An Edwardian postcard: the beginnings of organised tourism!

we shall see that the University was able to make a Research and Experimental Area exactly as he had advocated.

Although Lynch's reputation as a horticulturalist is widely recognised, it does not seem to be generally realised, even in Cambridge, that he had a strong interest in the British flora, and did a great deal of collecting himself. Evidence of this interest can be found increasingly in the Annual Reports from 1894 onwards, and the Report dated 7 June 1904 devotes a special paragraph to British plants which are said to 'have received as usual, a large amount of attention'. Like many Victorian field botanists, Lynch was interested mainly in the unusual variant or the rare species, but his enthusiasm to grow recognisably different plants meant that he accumulated great experience of the variation of plants both in nature and in cultivation, and indeed he gave a paper to the Royal Horticultural Society in 1900 entitled *The Evolution of Plants, illustrated by the Cultivated Nature of Gardens*, which is full of detailed examples from his own experience.

The interest taken by the Curator in the acquisition and cultivation of British plants seems not to have been strongly supported by the members of staff of the Department of Botany under Marshall Ward, and there is hardly any record in the successive Annual Reports of the period that Lynch received British plants for the Garden from any local University botanist. Rather we find that local field botany was being continued by amateurs like Arthur Bennett and E. W. Hunnybun, both of whom are mentioned as donors of British plants. One ought not to conclude from this that there was no interest in the British flora amongst the staff of Marshall Ward's Department, for the Professor himself was certainly interested, especially in the native trees and shrubs, as we shall see; rather it must have been that the 'new Botany' presented wider horizons and glittering prizes in research, and the tradition was temporarily broken in favour of the more exciting new fields.

In the next chapter we look at the impact of new subjects on the Department and the Garden during the early years of the present century, and in the final chapter how the rift between the field botanist and the University professional began to heal.

7

The New Botany School: Marshall Ward and his successors

Figure 61 The main gates on Trumpington Road.

The New Professors

By the dawn of the present century, plans were in existence for a group of new university buildings adjacent to, but on the opposite side of Downing Street from, the science buildings which had increasingly taken over the abandoned Walkerian Garden. From the outside only one trace of the Garden remained – the fine wrought-iron gates in Downing Street, which were removed and re-erected in their present position as the Trumpington Road entrance gates in 1909. (Inside the 'New Museums Site' one survivor of the Walkerian Garden could be seen until 1932: an old tree of *Sophora japonica*, cut down when the present Zoology Department building was erected.) Amongst this group of buildings, formally opened by King Edward VII and Queen Alexandra on 1 March 1904, was the present Botany School. At the time, this was generally agreed to be the best university botanical laboratory in Britain, and its establishment and success owed a great deal to the extraordinary scientific and administrative ability of Harry Marshall Ward, who succeeded Babington as Professor of Botany in 1895.

Marshall Ward was a product of the new popular education in experimental science, as different a man from his immediate predecessor as one could find. Born in Hereford in 1854 in a family of limited means, he was educated at the Cathedral School at Lincoln, and then at a private school in Nottingham which he left at the age of 14 years. He continued his education in evening classes organised under the Science and Art Department, with the aim of becoming a teacher of science. In 1874 he was admitted to Huxley's course of instruction in biology designed for teachers in training, and in the following year attended the classes run by Thiselton-Dyer and Vines in South Kensington to which we have referred in the previous chapter. He was obviously a student of quite exceptional ability, and Thiselton-Dyer, recognising his merit, helped his career in various ways, at first by using him as demonstrator in

Figure 62 Lime Tree (*Tilia × europaea*) planted by the main gate to commemorate the opening of the 'New Garden' in 1846.

the South Kensington course. No doubt encouraged by Vines, who was already a Fellow at Christ's College, Ward entered for the College, and obtained an Open Scholarship there, beginning his studies in Cambridge in October 1876. In 1879 he obtained first-class Honours in Botany in the Natural Sciences Tripos, and published his first paper, on the embryo-sac of Angiosperms, in the following year.

This is not the place to describe Ward's career in detail, but certain experiences and influences which shaped that career are very relevant to the modern development of Cambridge botany and therefore to our theme. The first of these, for Ward as for all the keen young botanists of that period, was a personal knowledge of German science. Like Vines, Ward worked with Sachs at Würzburg, where his practical ability, his skill in drawing and his meticulous attention to detail greatly pleased the master. These qualities in German science – indeed in German scholarship in general – were injected into British scientific education at that time, and survive strongly in the Cambridge botany course to the present day. We still expect students in the first year 'Biology of Organisms' practical classes to observe and draw carefully and accurately what they see and not what they are told they ought to see, and the practical examination still tests a range of qualities and abilities which are quite different from, and indeed often show very little correlation with, those 'book-learning' qualities tested in the written papers. Whilst it is obvious that there is a direct continuity in this element of our teaching back to the Marshall Ward era in the new building, it is also true, as we saw in chapter 5, that it was precisely those qualities of accurate observation and careful drawing which Henslow imparted in his 'new science' one hundred and fifty years ago. To that extent, Ward's influence was exactly in line with what Henslow had campaigned for in the University too early in the century.

In one important way, however, the opportunities for exercising these qualities were incomparably greater and more stimulating for Ward and his contemporaries than for Henslow. This was because of the development of the compound microscope, which opened the door to new worlds of biology, and gave to the students of that time a feeling that limitless possibilities were opening up for both pure and applied research. Ward's first job, no doubt arranged by Thiselton-Dyer, sent him out as 'Government Cryptogamist' for two years to study the devastating new leaf disease in the coffee plantations of Ceylon, and so began his eminent career as a mycologist and plant pathologist, from which the modern group within the Botany School takes its shape in a direct line of development associated particularly with the name of F. T. Brooks, Professor of Botany from 1936 to 1947.

Figure 63   The old *Sophora* tree on the Old Botanic Garden site, 1931.

The experience of tropical vegetation in Ceylon gave Ward an interest in trees and forestry, which led eventually to his appointment as Professor of Botany in the Forestry Institute of the Royal Indian Engineering College in Cooper's Hill, London; and it was from this Chair that he came back to Cambridge in 1895. On paper, therefore, Ward was at least as much a forester as a mycologist, but his writings on forestry and arboriculture are relatively little remembered today, and the six-volume work on *Trees* which he was writing for Shipley's *Cambridge Biological Handbooks* series was unfinished at his death with only four volumes published. One of the most appealing of Ward's numerous papers is that published in the *New Review* for 1891 entitled *A Model City; or, Reformed London*, where he makes a strong plea for more imaginative and experimental planting of exotic trees, shrubs and herbaceous plants in parks and gardens in English cities. The paper finishes with a sentiment which landscape architects might well echo today: 'it is perhaps a bold attitude to adopt, to deny that any fool can plant a tree, but it is certainly not true that any fool can plant it successfully; and when it comes to the matter of selecting what trees shall be grown in large towns, where and how they shall be planted and when and how they shall be pruned or otherwise treated, the requirements are far beyond those to be obtained from a fool. The moral of which may be read to be – plant more varied and better trees in cities and large towns, but take care they are properly planted and cared for.'

As with forestry, so with agriculture; Ward's views were strongly that the new experimental science should influence these enormously important practical subjects, both of which were developing for the first time at the university level. His interest in plant diseases of course overlapped into both areas of study, and it is characteristic of Ward that he wrote, almost in passing, a very useful little text-book on *Grasses* (1901) for university students who needed an up-to-date account of the common and important agricultural species. Of all his predecessors in the Chair of Botany in Cambridge, surely Richard Bradley would have been the one with whom Ward would have felt most sympathy – though there is no evidence that a nostalgic view of history had any attraction for him, and his death in 1906 at the early age of 52 years robbed him of any old age in which to reminisce or delve into the history of botany.

One of the very happy by-products of Ward's appointment at the Indian Forestry Institute in London was that his son Francis, who had been born in Manchester in 1885 just before the family moved to London, must have been brought up in an atmosphere of 'travellers' tales' from botanists and foresters returning from the East. This early influence was to form the dominant passion of his life; Frank Kingdon Ward became one of the most famous and successful of the great plant collectors and explorers of the Sino-Himalayan regions, and there is probably no large garden in Britain today which does not grow plants first introduced by him. When in Cambridge we admire *Meconopsis betonicifolia*, the magnificent Himalayan Blue Poppy, flowering in the Woodland Garden, or introduce students to that interesting hardy shrub *Acanthopanax wardii* (telling them that, however improbable it may seem, it is related to the ivy!), we are using plants which commemorate Kingdon Ward the son and, vicariously, Marshall Ward the father.

The excellent little handbook written by J. W. Clark, University Registrary, and published for the 1904 meeting in Cambridge of the British Association, gives a detailed account of the facilities in Marshall Ward's new Botany School in the year in which it was formally opened. It is interesting to find that most of the facilities are recognisably similar to those of the present day. For example, we find listed: a 'Morphological Laboratory for the practical study of advanced Morphology and Anatomy of plants', a 'Chemical Laboratory for experimental study and research in the chemistry of plants', the 'Professor's Laboratory and rooms for research in mycology and plant diseases', and the 'Laboratory for Plant Physiology . . . very fully equipped with apparatus for the experimental study of the physiology of plants'. To this we can add the Herbarium, the Library, the Elementary Laboratory and the Lecture Room (all described in the British Association Handbook, the last two exactly as they are today). Of course there are now entirely new activities quite unknown to Ward – biophysics, electron microscopy, palynology and others – but the main building and its sub-divisions remain basically as they were.

What do we know about the shape of the courses offered in Parts I and II of the Tripos for botanists at this period? The available staff consisted in 1904 of the Professor himself, who gave 'the General Course of Botany' and also an advanced course on Fungi; the Reader in Botany, Francis Darwin, who gave a course on Experimental Physiology; the two Lecturers in Botany, Seward (who succeeded Marshall Ward as Professor) and F. F. Blackman, founder of the modern plant physiology sub-department; and two University Demonstrators, Hill and Gregory. The Professor's General Course consisted of the following parts: an introduction to the study of organography, morphology and anatomy, and physiology of plants in the Michaelmas Term; an 'Evolutionary Course on the Biology and Classification of the Fungi, Algae, Bryophyta, Vascular Cryptogams and Conifers' in the Lent Term; and a course on the Systematic Botany of the Flowering Plants during the Easter Term. This last course, we are told, is 'supplemented by visits to

Figure 64 Marshall Ward's Elementary Botany class in 1906. The Laboratory is still in use in the same way today.

the Botanic Garden, and by excursions into the surrounding country during the Long Vacation, for the study of the Flora of the district'. The tradition that the Professor himself should take a personal interest in the elementary teaching, and give at least some of the lectures himself, a tradition so strongly reinforced by Ward's successor Seward, has remained a strong element in the Botany School instruction right down to the present day.

Before we leave Marshall Ward's achievements we might note one very successful and permanent piece of cooperation between him and Lynch. Trained up the south wall of the Botany School like 'espalier' pears, we have, to the present day, a pair of maidenhair trees, *Ginkgo biloba*, male and female, which must be unique. At the base of each was until recently a metal label of the 'Lynch Botanic Garden' type which records that the plants came as scions for grafting from Montpellier in 1896, and the Annual Report of the Botanic Garden Syndicate for 1897 records 'male and female *Ginkgo biloba* (grafts successfully established on seedling plants)'. Although no record of the actual planting has yet emerged, there seems to be no doubt that Lynch made the grafts, raised a male and a female plant in the Garden nursery and planted them against the new building . . . an appropriate symbol of the happy cooperation between Professor and Garden in the provision of living plant material. To the present day it is accepted without question by Garden and Department alike that the annual autumn pruning of the Ginkgos shall be done by Botanic Garden staff, not by the gardeners of the Estate Management Service who are responsible for all other maintenance on the Downing Site.

Ward was succeeded in 1906 by Albert Charles Seward, a man of considerable gifts and phenomenal energy, who was able to combine to a very unusual degree efficient administration, highly successful teaching, and very effective research. Born in Lancaster in 1863, he was keenly interested in natural history as a boy, and came from Lancaster Grammar School to St John's College where, contrary to his parents' desire that he should enter the Church, he turned to Natural Science. By 1890 he was appointed to a Lectureship in Botany, and became a Tutor of Emmanuel College in 1900. His Collegiate career was crowned by his election as Master of Downing College in 1915, and his dual service to College and University was finally marked by his election as Vice-Chancellor in 1924, a post he held for the customary two years. He was knighted in the same year, 1936, that he retired from the Chair.

One of his many eminent pupils, Professor T. M. Harris, wrote in an obituary notice in 1941 of the qualities of Seward's scientific research and teaching. On his research Harris says: 'it is most difficult to distinguish periods in his scientific career, for at all times

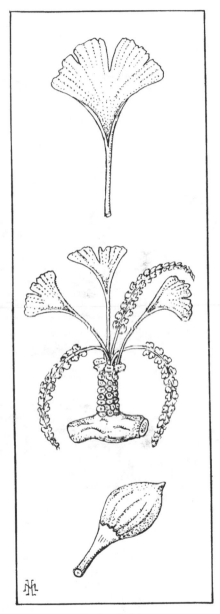

Figure 65   *Ginkgo biloba.*

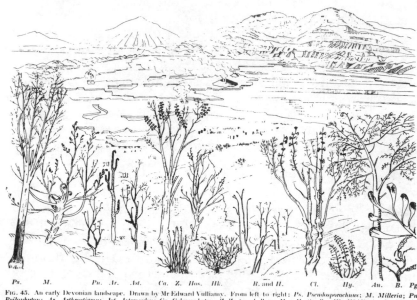

FIG. 45. An early Devonian landscape. Drawn by Mr Edward Vulliamy. From left to right: Ps. *Pseudosporochnus*; *M. Milleria*; *Pn. Psilophyton*; *Ar. Arthrostigma*; *Ast. Asteroxylon*; *Ca. Calamophyton*; *Z. Zosterophyllum*; *Hos. Hostimella*; *Hk. Hicklingia*; *R. Rhynia*; *H. Hornea*; *Cl. Cladoxylon*; *Hy. Hyenia*; *An. Aneurophyton*; *B. Broggeria*; *Pg. Psygmophyllum*. The reconstruction of these plants is based mainly on drawings and descriptions by Kidston, Lang, Kräusel and Weyland.

Figure 67   Illustration by Vulliamy from Seward's *Plant Life Through the Ages*.

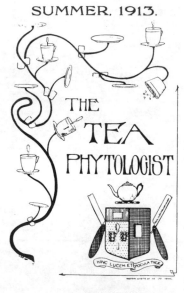

Figure 66   Cover of the third number of the *Tea Phytologist*, a light-hearted product of the New Botany School, which still survives as an 'occasional periodical'.

his interests were so wide as to include the whole of palaeobotany and much of recent botany too . . . Throughout his writings his preferences for certain sides are clear; he liked things that were open and large, and his efforts were directed to making them more open and larger'. On his teaching, we are told: 'Seward was a splendid teacher and a most successful head of the largest Botany School in the country. He thoroughly enjoyed lecturing at all levels, but I believe he liked the elementary lectures best, and certainly it was a joy to hear them . . . His general writings reveal him as a teacher, taking this word in its widest sense. He had no use for science which is the possession of the select few, but believed in it and worked for its dissemination'. His achievements reflect both his broad view and his phenomenal capacity: he was elected a Fellow of the Royal Society at the early age of 35 years, and later became Vice-President and Foreign Secretary; he served also as President of the Geological Society, of the Botanical Section of the British Association, and of the Fifth International Botanical Congress. Under his sensible and efficient rule the Botany School of Marshall Ward expanded in influence both as a teaching and a research Department. His semi-popular text book of evolutionary botany entitled *Plant Life through the Ages*, just half a century old, can still be read with pleasure and profit today.

The new branches of botany

Bower (1938), recalling botanical teaching and research in Cambridge around the turn of the century, has an attractive simile to describe the rapid diversification. He writes: 'the development of botanical study as a whole might be compared with the progress of a flock of sheep, advancing fanlike over a large plain, but with irregularities of formation defined either by individual enterprise or by the varying richness of the pasture; or by both'. The local concentrations of sheep – to continue the simile – were to prove more permanent in some cases than in others and, as we have seen, several are clearly visible right to the present day.

Figure 68   Frontispiece from Elwes and Henry's book on trees.

Of those which made no more than a temporary impact on the Cambridge scene, the study of forestry has a special interest, because of the consistent tradition of the Garden as an arboretum and because of Marshall Ward's own special knowledge and experience. In 1907, soon after Ward's death, the University instituted a Readership in Forestry, and appointed to it Augustine Henry, a gifted amateur plantsman and explorer of the Far East who was collaborating with H. J. Elwes on the monumental seven-volume work, *The Trees of Great Britain and Ireland*, published between 1906 and 1913. Henry used the Botanic Garden for raising families of seedling trees, and in return supplied interesting and rare additions to the collections. His paper *On Elm seedlings showing Mendelian results*, published in 1910, incorporates data on large families of seedlings raised from seed from local elms partly in the Botanic Garden, some of which were planted out on 'the experimental forestry plot on the University Farm at Cambridge'. Henry is also responsible for the description of the plane tree *Platanus cantabrigiensis*, based upon a single tree which still grows in the Botanic Garden. (This variant of the familiar London plane was discussed and illustrated by Lynch in the last paper he wrote in 1924, the year he died.) Henry moved to the Chair of Forestry in Dublin just before the outbreak of the War in 1913, a post from which he retired in 1926. The Garden still houses a fine, heavy table made of a single piece of English elm (*Ulmus procera*) which was presented to the Forestry School by Elwes from his estate, Colesbourne Manor, Gloucestershire, about 1921; it came to us when the Department of Forestry was dissolved in 1932 and Dr A. S. Watt, Lecturer in Forestry, joined the Botany School staff.

Figure 69   *Platanus cantabrigiensis* Henry, drawn from the type tree still growing in the Garden.

Turning to the permanent new subjects, the one which had the most spectacular development on an international scale in the first decade of the century was undoubtedly genetics. We have already seen how, locally, the Garden played an important role in the very early days of the science by providing facilities for growing the families of experimental plants studied by Bateson, Saunders and others. By 1906, when Seward was appointed to the Chair, the increasing demand for space under glass to grow families of experimental plants for genetical research was such that the Syndicate sanctioned the construction of 'a new pit, 52 feet in length, uniform in plan with the three existing pits'. This was the first fixed provision in the Garden for experimental plants. Successive Annual Reports reveal an increasing concern – which is presumably being expressed by the Curator through the medium of the Report – that the sheer size of the growing demand from genetics was proving too much. The opening paragraph of the Report dated 7 June 1909 is ominously headed: 'Experiments on Plant-breeding and Increased Pressure on Available Space', and the paragraph reads as follows:

The numerous experiments on Plant-breeding referred to in the last Report have been continued. The demands upon the resources of the Garden arising from these have increased and it is known that more space under glass than ever before will be required during the coming winter. Although further accommodation should be provided, it is hoped that by some sacrifice in the working of the Garden the necessities of this important work will be practically met for the present. The sacrifice here referred to involves the danger of decreased efficiency in the provision of specimens for teaching purposes at the Botany School, the space hitherto devoted to the cultivation of specimens for class use being unavoidably lessened. Increase of space is also necessary out-of-doors for special cultures and it may be necessary to devote the next convenient allotment that falls vacant to the use of the Botanic Garden. It is further hoped that an arrangement may be made with one of the colleges for ground upon which to plant trees and shrubs for which there is no provision in the Botanic Garden.

In 1908 Bateson's eminence as a geneticist was recognised by the University, who created for him a Chair interestingly enough first designated as a Chair of Biology; but by 1910 he had moved to the John Innes Horticultural Institute, and R. C. Punnett succeeded him in Cambridge. Both Bateson and Punnett served as Syndics of the Botanic Garden, and were obviously pressing for greater use of the 'Field' allotments for experiment. In 1913 we read: 'during the past year four allotments have been rented by the Professor of Genetics and breeding experiments have there been carried out by a number of investigators'. In 1914, however, the new Genetics Institute (later the Department of Genetics) was installed in Whittinghame Lodge, Storey's Way, where it developed its own facilities for growing experimental plants and animals, and in any case the outbreak of war meant that the number of research projects needing space in the Garden was gradually reduced until 1919.

Neither Bateson nor his pupil and successor Punnett were sympathetic to the growth of cytology (indeed, Bateson never really accepted the chromosome theory of heredity), and much credit is due to F. T. Brooks, Lecturer in the Botany School and later Professor, who saw in the immediate post-war years the interest and relevance of chromosome studies and included them in his teaching. Important work such as that of Miss Saunders on the Garden Stock *Matthiola* continued to be catered for in the Garden on allotments and in the 'experimental pit'. The outstanding post-war use of the experimental facilities took place between 1922 and 1929, when C. C. Hurst conducted his very extensive cytogenetic investigation of the genus *Rosa*. From these studies arose, as permanent memorials to Hurst, roses named after the Garden, one of which, the hybrid between the Himalayan *R. sericea* and the Chinese *R. hugonis*, is the well-known *R.* 'Cantabrigiensis', (now called *R.* × *pteragonis* 'Cantabrigiensis'), which won the Royal Horticultural Society's Award of Merit and the Cory Cup (for the best new hardy hybrid shrub) in 1931 (see Frontispiece).

Continuing Bower's simile, we might liken the rapid growth of genetics in this century to a 'local concentration of sheep' which soon became a separate flock in a new, rich pasture. The growth of ecology looks very different, however, being much more a product of 'individual enterprise', for there seems to be no very clear reason why the study of plant communities and inter-relations should not have developed much earlier in the history of the science. The 'individual enterprise' responsible for the development of plant ecology in Britain is undoubtedly A. G. Tansley, who entered Trinity College in 1890 and obtained first-class Honours in Part II Botany in 1894. His early research and teaching experience was mainly in University College, London, from which he returned to a Lectureship in the Cambridge Botany School in 1906. The excellent paper by his eminent pupil, Sir Harry Godwin (1977) which was delivered in 1976 as the first Tansley Memorial Lecture, reveals what peculiar attributes of enthusiasm and organisational ability the young Tansley brought to bear on the problem of establishing ecology as a discipline in British universities. When we recall that Tansley was responsible, largely single-handed, for founding the *New Phytologist* (which he edited for 30 years), the British Ecological Society (of which he was first President), and even the International Phytogeographic Excursion (I.P.E.), all before he was 45 years old, we have some idea of the stature of this 'man of ideas', a philosopher rather than a scientist of the familiar nineteenth-century mould.

Tansley in Cambridge University and the Cambridge Botany School is more difficult to assess. Godwin speaks of the period from 1914 to 1927, when Tansley took the Chair of Botany in Oxford, as

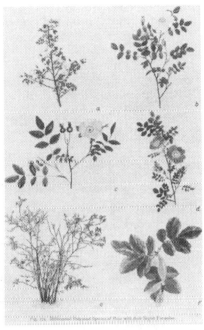

Figure 70 Plate from Hurst's account of the genetic history of Roses, published in 1925.

one of 'disillusion and disaffection' for him. No doubt the War, with its destructive and dislocating effects, was partly responsible; but it must also have been true that there was some opposition to any radical change in the balance of teaching in the university courses. The difficulty was no longer the 'dead Linnaean taxonomy', of course, which as we have seen had been reduced to a single Easter Term Course in Marshall Ward's time. Indeed, some thought that the dethronement of taxonomy had gone too far, in that a knowledge of the British flora, so necessary to descriptive ecology, had been an unfortunate casualty of the post-Babington reaction. Certainly part of the problem concerned the attitude of experimental scientists to what they were prone to stigmatise as mere observation and cataloguing. As Tansley saw very clearly, there are inevitably two stages in the development of ecology, an initial descriptive and taxonomic phase, followed by an analytical and experimental one, and although all description need not be complete before any experiment, some agreed body of information is a necessary pre-requisite for the development of quantitative and experimental ecology. Tansley himself admitted to two weaknesses, both of which may explain part of his difficulty: he was in fact rather poorly equipped in knowledge of the British flora, and he had no training in, and indeed little aptitude for, experimental science. But the foundation he laid was so effective that, through his pupil Harry Godwin who eventually held the Cambridge Chair, and the generations of students who studied in Cambridge from the thirties onwards, both these limitations were completely removed.

Tansley, brought back from London in 1906 by Seward, was elected on to the Botanic Garden Syndicate in 1908, and acted as Secretary from 1910 to 1915. The first explicit reference to the use of the Garden for ecological research comes in the Report dated 3 June 1912, in which the following sentence appears: 'In the reserve ground at the back of the range of plant-houses new experiments in Ecology by Mr Adamson, Miss Hume and Miss Pallis have recently commenced, and special facilities have been provided in each case.' One of these investigations, that originally undertaken by Miss E. M. Hume at Tansley's suggestion, concerned competition between the two British species of *Galium, G. saxatile* a calcifuge species, and *G. sterneri* (*G. sylvestre*) restricted to limestone. The experiments were taken over in 1913 by A. S. Marsh, but he joined the army on the outbreak of war and was killed in 1916. Tansley continued the observations on the experimental cultures set up by Miss Hume and Marsh, and reported on the results in a paper published in 1917 entitled: *On competition between* Galium saxatile . . . *and* G. sylvestre *on different types of soil.* Much earlier than this, however, in 1900, Lynch had made in the Garden 'new

Figure 71   The dwarf Pine, *Pinus sylvestris* 'Moseri', on the Old Rock Garden, *c.* 1927. This tree still grows there today on what is now the Terrace Garden.

beds . . . to illustrate the vegetation characteristic of different kind of soil, with special reference to plants met with in Chalk districts and at the Seaside', and by 1908 he was referring to these as the 'Ecological Beds'. These beds survived in a modified form until the early 1960s; their last trace, still flourishing, is in the old box tree (*Buxus sempervirens*) which stands immediately west of the Terrace Garden and south of the Succulent House. How far Tansley cooperated with Lynch in the design of the 'ecological beds' is not known; certainly Lynch had sufficient knowledge of the British flora and vegetation to have designed and stocked the beds without assistance from any member of the Botany School staff. Perhaps C. E. Moss, Curator of the Herbarium from 1907 to 1916 and author of the uncompleted *Cambridge British Flora* (1914–20), who was a pioneer ecologist as well as a taxonomist, assisted and encouraged Lynch in such enterprises. Moss's departure for South Africa in 1916, where he became Professor of Botany in Johannesburg, and died at the early age of 58 years, was a loss to Cambridge botany, although, as we shall see, his student Humphrey Gilbert-Carter was to inaugurate a new era in the study of systematic botany in the post-war period.

The final 'new' branch of Botany which we must briefly mention, at least so that its historical links can be seen, is agricultural botany. Here, after nearly 200 years, we find the University accepting and catering for a scientific approach to crop cultivation for which the first Professor of Botany, Richard Bradley, had hoped in vain. No longer was it held against a botanist that 'he knew no Latin and Greek' (though some minimum knowledge of the former remained an essential University entrance requirement for many years to come), and the exciting atmosphere of applied research in Marshall Ward's laboratory was the ideal training ground for the first Professor of Agricultural Botany, R. H. Biffen. In the words of a later Professor of Agriculture, Sir Frank Engledow: 'Imaginative and gifted in experimentation, [Biffen] was caught up in the enthusiastic study of heredity kindled by the revelation to the scientific world in 1900 of Mendel's laws . . . Biffen perceived, with clarity and conviction far beyond contemporary opinion, the powerful agency offered by this new understanding of heredity, to the improvement of cultivated plants by hybridising.' In 1910 was opened the new building for the School of Agriculture south of the Botany School, and in 1912 the Government financed the Plant Breeding Institute, with Biffen as its first director, beginning the work which has continued, with international fame, to the present day.

By the time that the University was struggling to return to full vigour after the end of the First World War, it was no longer easy to see the many activities within the Botany School created by

Marshall Ward as logically related or effectively integrated. Tansley expressed this difficulty very clearly in his Presidential Address to Section K (Botany) of the British Association in Liverpool in 1923, in a remarkable paper which hardly mentions ecology, dealing instead with the distressingly rigid separation between phylogenetically-obsessed morphology, and physico-chemical plant physiology. He ends with the following plea for a broadly-based elementary botany dealing with form and function together:

If botany, as the science of plants, is to retain any meaning as a whole, somebody must retain the power of looking at it as a whole. And if, as teachers, we fail to keep touch with the newer developments, and are consequently no longer able to focus the whole subject from a viewpoint determined by current knowledge, this power will come to be possessed by fewer and fewer botanists, and the subject will definitely and finally break up into a number of specialised and unco-ordinated pursuits. Do we want that to happen? I think that most botanists would answer "No!" I do not think there can be any question that the most advanced research worker, as well as the student who never goes on to research, benefits substantially by having had a training which is at once the broadest and the most vital that is possible. As science continuously advances and necessarily specialises, the unexplored fields which lie between the traditional lines of research become of more and more relative importance. They cannot receive adequate attention – the student can, indeed, hardly become aware of their existence – unless his introduction to the subject is continuously informed by the widest outlook and the clearest apprehension of the essential relations of the phenomena of plant life.

Whole-plant botany and the modern Botanic Garden

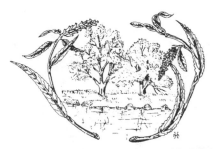

Figure 72 The River Cam with its willow trees, favourite teaching material of Gilbert-Carter.

Humphrey Gilbert-Carter, first Director of the Garden

When Lynch offered his resignation in 1919, at the age of 69 years, the Botanic Garden Syndicate were provided with an opportunity, which they seized, to alter the mode of direction of the Garden towards scientific botany and away from horticulture. In the Annual Report for 1920 their intention is expressed as follows: 'They proposed to abolish the office of Curator and to substitute for it a Directorship, to be held by a trained systematic and economic botanist, who would take part in the teaching and training of university students, and in addition to appoint a practical gardener as Superintendent.'

The post of Director was advertised, and several candidates were considered. The Senior Foreman under Lynch, Mr F. G. Preston, was appointed Superintendent, and was already in office when the University appointed as first Director of the Garden Humphrey Gilbert-Carter, who returned from Calcutta in March 1921 to take up his post.

The new Director was 37 years of age at the time of his appointment, already married, and with considerable experience of tropical botany from his service in India as Economic Botanist to the Botanical Survey. He was born in 1884, the second son of Sir Gilbert Thomas Gilbert-Carter, Governor of Lagos (now Nigeria); he went to Tonbridge School and from there to Edinburgh University where he read medicine. After post-graduate study in Marburg, he turned to botany as a career and as we have seen came to work in Cambridge as an advanced student under C. E. Moss in 1909. From that period began a lifelong friendship between Gilbert-Carter and Tansley which meant a great deal to both men, resulting among other things in the publication in the 1930s by the Oxford University Press of three of Gilbert-Carter's most important books, namely: *Our Catkin-Bearing Plants* (1930, 2nd edn 1932), the translation of Raunkiaer's Danish work on *The Life Forms of Plants* (1934, 1937), and *British Trees and Shrubs* (1936).

Figure 73 Humphrey Gilbert-Carter at the time of his appointment as Director of the Garden in 1921.

Gilbert-Carter held the new post of Director in conjunction with the Curatorship of the Herbarium in the Botany School for the first nine years, and after that with a Lectureship in Botany. In this way he was intimately associated with the teaching in the Department, and students who were interested in systematic botany came to look upon the resources of the Garden as an important part of their botanical studies. The expressed intention of the Syndicate was happily realised. It was, however, the unusual personal qualities of the Director which made his influence on the shape of Cambridge botany so important in the difficult inter-war years. As Professor Clapham put it in the memorial volume to Gilbert-Carter (Gilmour & Walters (eds.), 1975), recalling the Part II lectures he received: 'These were certainly talks rather than lectures, and a series of more or less disconnected talks . . . They were given in the small room at the Botanic Garden, and the subject of the talk on any particular day was dictated partly by what flowers happened to be available on that day but mainly, it seemed to us, by the whim of the teacher.' Such idiosyncratic lecturing inevitably divided his audience into those who were fascinated and delighted by it and those whom it bored or offended. With informal teaching in the Garden, or excursions to visit local plants and local 'pubs', and at memorable tea-parties in the Director's house, Gilbert-Carter influenced and in a real sense educated many of the leading botanists of the present generation, and this influence is still widely felt. To take a single familiar example: the new British Flora of Clapham, Tutin & Warburg (1952) was a direct product of the Gilbert-Carter teaching, for not only were all three authors pupils of his but, as Professor Tutin relates, the very idea of writing the Flora was the outcome of a 'Humphrey walk': 'On a winter afternoon shortly after the [Second World] war, we walked together through Trumpington to Grantchester and, as we were about to come back across Grantchester Meadows, he suggested that we might see if Tansley was in. He was, and on his own, so he asked us to stay for tea. Over tea Tansley asked me to write a British Flora. I had never thought of such an undertaking, but before we left plans had been drawn up and were speedily put into action. So Humphrey not only instructed the three of us in field and herbarium, but was a vital link in the chain which led to the production of the Flora.'

What were the qualities which Gilbert-Carter used so effectively in his teaching? Listing them is not easy, but one at least is quite straightforward. He was a plantsman who enjoyed the game of getting to know new species and, even more, of visiting again 'old friends' on field excursions. The mantle of Ray fitted him comfortably, and he made the local excursion in the Cambridge area once again a delightful, educative experience for generations of students

Figure 74   Sunday tea-party at Cory Lodge about 1935.

Figure 75   Student excursion at Swaffham Prior, 1942. Gilbert-Carter in the foreground.

after the First World War. Trees were his great love, and here again he fitted so well into a strong Cambridge tradition, which Henslow had founded and Marshall Ward and others had continued. We recall with pleasure the words with which he addressed his first meeting with the Botanic Garden Syndicate in 1921, recommending the planting of good specimen trees: 'Let us look ahead and think of those who will take our places. Neither shall we ourselves die altogether if the coming generations of students remember us when they study the great trees on the lawn, or the elders of the University and town bless our memories as they rest beneath their shade.'

A second quality was his wide international view of botany shared, of course, with Tansley, a view at many different levels. Like most science students of his generation, he had learnt to value German scholarship; his particular contribution was to introduce into the Cambridge teaching of systematic botany a knowledge of the Englerian system and to provide as early as 1913 a small English text-book, *Genera of British Plants*, which made that system available for the first time to British students. The adoption of the (up-dated) Englerian arrangement in the great cooperative work just completed, *Flora Europaea* (1962–1980), is a decision which stems directly from Gilbert-Carter's 'European view', so effectively transmitted to his students during the inter-war years. Moreover, his broad outlook took in both tropical and economic botany, where his Indian experience gave him a richly valuable background; he had an endearing tendency to assume that all his Part II students of systematic botany were destined to find themselves in after years struggling to identify the rich floras of the tropics in colonial forestry or agricultural posts.

Undoubtedly, however, it was his extraordinary linguistic accomplishments which imparted to his teaching and to his writings that unique quality which captivated his audience. His generation was still 'well educated' in the public schools, with a thorough knowledge of Latin and Greek, and on this foundation he built a remarkable structure, which began with the more important Western European languages, including fluent Danish, to which he added a range of Eastern languages – Hindi, Urdu, Persian and Arabic. The *Descriptive Labels for Botanic Gardens*, which were reprinted in the memorial volume, contain many impressive examples of Gilbert-Carter's love of plants and the words associated with them. Some of these labels printed on special waxed card are still in use in the Garden today. In his retirement Gilbert-Carter continued to learn new languages for pleasure, though he complained that Russian was too much for him at the age of 75 years.

The Garden in 1921, thanks to the ordered régime of Lynch and the efficient administration of Seward, was basically a well-organised place, although war-time shortages and difficulties had left their mark, and a return to normality was not made easier by a general shortage of money experienced by all University departments. There is a rather gloomy paragraph on 'Finance' prominently placed in the Annual Report dated 7 November 1921, which points out that wages for gardeners were more than double what they were in 1914, and that, given such trends in costs, 'it will be impossible to maintain the efficiency of the Garden'. The same Report records the first donation by 'Mr Reginald Cory (formerly of Trinity College) of Duffryn, near Cardiff', who gave £1000 to employ extra labour and buy materials to make up for war-time neglect. This joint theme of financial difficulty and Cory's generosity became regular features of the successive Annual Reports of the 1920s. In 1925 the new house for the Director was built to the design of Baillie Scott, and called Cory Lodge to commemorate Cory's donations; the Director moved in with his wife and family in the autumn of that year, and the house has been continuously occupied by successive Directors ever since.

Under the new arrangement, all the day-to-day administration of the Garden was the responsibility of the Superintendent. The Director was responsible for carrying out the policy of the Syndicate, for the teaching of Systematic Botany in the Garden under the Professor of Botany and in other ways for the furtherance of 'the interests of botanical teaching in the University'. More specifically, the Syndicate laid down that he should deal with the Garden correspondence, that he should receive 'distinguished visitors' to the Garden and that he should be responsible for the identification, labelling and description of the collections. Both he and the Superintendent were asked 'to make whenever possible a daily round of inspection of the Garden in furtherance of the duties of supervision allotted to each'. The new Director took all these duties seriously; he enjoyed a routine which normally consisted of a visit to the office to deal with his correspondence, followed by an appearance in the Botany School Herbarium from which he regularly adjourned with colleagues (by no means exclusively botanists) to the adjacent *Bun Shop* for a glass of beer before lunch. After lunch, followed, at least in his later years, by a siesta, he did his daily inspection of the Garden before tea, and any offending plant label incorrectly worded or placed would be ostentatiously turned round – a sign by which the Garden staff would know that they must consult the Director himself to correct the fault. This personal concern with the accuracy of the labelling, combined with a real love of the plants, set a standard for which the Garden has been and still is quite famous.

Viewing his appointment as one involving the preparation of

interpretative material for students using the Garden, Gilbert-Carter gave high priority to writing a *Guide* which was published in 1922 (with a second edition in 1947). This little book, like all his writings, is highly idiosyncratic, and the Introduction to the second edition records quite tersely how it came to be written. It was, Gilbert-Carter says, 'one of Reginald Cory's numerous benefactions to the Garden. It was published by his request, at his expense, and sold at a heavy loss'. He continues: 'When I first met Cory in 1921 he was already interested in the Botanic Garden. He asked me to write a *Guide* and I did so. It was not an ordinary book, and much of its contents have no wide appeal.' It is certainly no 'ordinary book', but the author may have been too modest in thinking that much of it has 'no wide appeal'. The purpose of the book is clear: it is not designed for the casual visitor whose appreciation of the Garden does not extend beyond its function as a restful and attractive public park, nor is it aimed at the ordinary gardener. It assumes some familiarity with systematic botany such as the university students of botany in the inter-war years could still be supposed to have from their school teaching. Thus, a description of the magnolias near Brookside (not yet part of the Garden even when the second edition of the *Guide* was written) runs as follows: 'Beyond the Holly tree at the end of the Border you will find the Magnolias. The family *Magnoliaceae* is allied to the *Ranunculaceae*, and shares with it numerous spirally arranged carpels. In *Magnolia* the axis on which these carpels are borne is remarkably long. The perianth is wholly petaloid and consists of three whorls. The allied genus *Liriodendron* (Tulip Tree) has the three outer perianth segments reflexed. The leaves of *Liriodendron* are so abruptly truncate as to appear to have had their ends cut off with scissors.'

In spite of Reginald Cory's generous help in the first post-war years, the financial problems of the Garden, perhaps reflecting the general economic depression, were worsening in the late 1920s. A mood of general pessimism hangs over the Minutes of the Executive Committee in 1927 and 1928, so much so that a sale by the University of a strip of allotment land to the Ortona Bus Company, enabling the Company (now Eastern Counties) to build the present bus garage on the eastern Hills Road boundary of the Garden, seems to have gone through without protest from Executive Committee or Syndicate. In January 1929 a special Syndicate was appointed 'to consider the organisation and finance of the Botanic Garden and the relations between the Garden and the Department of Botany and other scientific Departments'. The members of the Syndicate included Seward as Professor of Botany, his former student H. Hamshaw Thomas, who was now a Lecturer in the Department, and Tansley who was already Professor of Botany at

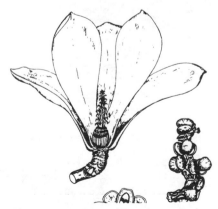

Figure 76 *Magnolia* plate from Hickey & King (1981) drawn from live material grown in the Garden.

Oxford. The Report, published in May 1929, is well-constructed and thorough, and reviews the history of the Garden and its role in teaching, research and amenity. The most important of its thirteen recommendations are that the Garden 'should become an integral part of the Department of Botany', that the Professor of Botany 'should be the responsible Head of the Garden', and that the finances of the Garden should be more formally and obviously a charge on the University because of its importance for teaching and research. The Syndicate would cease to be an executive body, and its future concern should 'be entirely with the amenities of the Garden'. A tentative recommendation about the allotment land was also included: 'That consideration should be given by the University to the fact that a part of the land adjoining the Garden could be sold under suitable restrictive conditions without detriment to the present and probable future needs of the Garden.' Happily we can report that such 'consideration' as the University gave to sale of the land in the years following the adoption of the Report was entirely negative; some approaches on behalf of both Heffer's Printing Works and the Bus Company for further sales or leases of adjacent allotment land were rejected, and the land remained in annual allotment tenancy until 1950, when the construction of the 'New Area' began.

From 1931, when the new Executive Committee of the Botanic Garden began to function, until after the Second World War, the Syndicate merely met annually to agree their Report. Their most pressing immediate concern was to launch a special Appeal on behalf of the Garden in 1932, which aimed at obtaining enough money to finish the rebuilding of the glasshouse range begun in the previous year, and to provide an endowment fund for future Garden maintenance. This Appeal was successful in that it attracted a gift of £6,000 from Mr W. J. Courtauld to finish the glasshouse re-building in teak. At the same time as the general appeal was launched, Professor Seward made a local Appeal for 'Friends of the Botanic Garden', a scheme which made available, for an annual subscription, a Sunday key to any Cambridge resident whether a senior member of the University or not.

In May 1934, Reginald Cory died suddenly, bequeathing the residue of his estate to the University for the benefit of the Botanic Garden. Four years before he had married Miss Rosa Kester, secretary in the Garden, who properly received the first share of the considerable fortune. The national publicity attendant on the publication of the will gave a misleading impression that the Garden would immediately benefit, and Seward was moved to write to *The Times* on 26 January 1935 explaining that this unfortunately was not the case. In fact, because of legal and financial complexities, the Cory Fund was not available to the University until the middle of

Figure 77   Old entrance to the Garden from Bateman Street, *c.* 1935. Until 1951 the office was situated here, in part of the Superintendent's House. The fine tree of *Xanthoceras sorbifolium*, seen here, was blown down in a gale in August 1968.

Figure 78   Cedar of Lebanon, *Cedrus libani*, in the middle of the Garden as laid out in 1846. In the background is the glasshouse range erected by Lynch in 1888–92, and rebuilt in 1934.

the Second World War, so that the financial limitations were strongly felt in the Garden throughout the whole inter-war period.

'Henslow's Garden' becomes a reality

The history of the Garden after the publication of Cory's will inevitably reads like a 'holding operation', an impression of course accentuated by the five years of the Second World War. The Garden was run with reasonable efficiency as a service institution to the Department and as a public amenity, although neither the Director nor the Superintendent was either able or willing to undertake any new schemes. The Victorian optimism of Lynch had quite evaporated, but what survived was remarkably robust and quite exceptionally attractive. In particular, the combination of Gilbert-Carter's love of trees, Preston's horticultural skill and Seward's very real interest meant that the woody collections were cared for

Figure 79   Vertical aerial photograph of the Garden taken on 26 August 1945, showing the main lawn with war-time vegetable plots, and the whole of the eastern half still divided into allotments.

Figure 80   *Cytisus battandieri*, a fine new horticultural shrub, first grown in the Garden in 1935.

and improved. The successive Annual Reports, which generally list interesting trees and shrubs received during the year, provide ample evidence for this. We can trace, for example, the development of the excellent collections of certain shrub genera such as *Berberis, Lonicera* and *Viburnum* for which the Garden is now widely known, and we can find recorded the early introduction of new horticultural shrubs which later achieved popularity, such as *Cytisus battandieri* introduced from the Atlas Mountains in North Africa in 1922, which we started to grow in 1935.

On 2 June 1943, in a very reduced Annual Report showing marks of war-time austerity, appears the following passage:

MR REGINALD R. CORY'S BEQUEST

The Syndicate learned with deep satisfaction and gratitude that this handsome bequest had now begun to accrue to the Garden (Reporter, 1942–43, p. 355). The bequest has proved to be much larger than anticipated, though only a small part of it is available for upkeep. After the war there should be great opportunities of extending the range of plants cultivated in the Garden and of generally enhancing its

beauty. The Managers of the Cory Fund have recently been
appointed, in accordance with the terms of the will.

This piece of good news is followed by two paragraphs recording
the termination by Professor Brooks (who had succeeded Seward
as Professor of Botany in 1936), of two schemes providing small
but useful financial help to the Garden: the 'Friends' scheme which
Seward had launched in 1932, and a remarkable Fund known as the
'Somerset Employment Fund', begun by a Mrs Somerset in 1900,
and originally designed to pay the wages of local unemployed who
were willing to work in the Garden. Supporters of the Garden
under both schemes were thanked for their practical help 'during a
difficult period'.

The 'Cory Managers' duly held their first meeting, with Brooks
as Secretary, in June 1943, a meeting at which a list of 39 sugges-
tions for spending the bequest was drawn up and given preliminary
consideration. It soon became apparent, however, that the rela-
tionship between the Syndicate, the Executive Committee (a Committee of the Faculty Board of Biology 'A') and the new Cory
Managers was in need of some careful thought, not least because the
priority for certain developments of scientific facilities in the
Garden might in the opinion of some at least be much greater than
that for schemes to improve the amenities. Such questions as the
future use of the allotment land and the establishment of a Research
and Experimental Area in the Garden clearly involved important
questions of priority. As a result of this concern, new recommendations on the direction and administration of the Garden were made
in a Report of the Council of the Senate dated 20 May 1946, and
were adopted by the University. This scheme, under which the
Garden is made, for the first time, a Sub-department of the Department of Botany, the 'new' Syndicate is responsible for determining
the general policy of the Garden, and the Director of the Garden
acts as the Secretary of the Syndicate, is essentially the structure we
have today. This new structure was to operate 'when the office of
Director next becomes vacant', which was in 1950 on the retirement of Humphrey Gilbert-Carter. The Report envisaged that,
with the possibilities of development under the Cory Bequest, 'the
responsibilities of the Director will be increased, and he is likely to
have greater facilities for scientific work': accordingly, the Ordi-
nances relating to the Director were changed, and his duties re-
defined to the present ones which lay emphasis on 'the care of the
Garden, the upkeep of the premises, the maintenance of the collections and the provision for teaching and research'.

The sense of new opportunities for the Garden was enhanced by
the approximate but not accurate coincidence of the retirement of

Superintendent and Director, which meant that R. W. Younger was appointed to succeed Preston as Superintendent in 1947, and was already installed and ready for the great developments on the appointment of my predecessor, J. S. L. Gilmour, as Director in 1951. The feeling of contrast with the 'old ways', and the excitement of post-war development with apparently unlimited money available, are very strong in the memories and writings of those who worked in the Garden or served on its committees in the 1950s. The new Office for the Garden in the handsome early Victorian house at 1 Brookside, purchased with the Cory Fund together with 47 Bateman Street and Brooklands Lodge, provided an ideal setting from which the new Director could plan great changes and extensions for the Superintendent and his rapidly-growing staff to carry out. In a touching 'reminiscence' in the Gilbert-Carter memorial volume, the Superintendent paints the following picture of the Garden and its Director in 1947:

At that time mechanisation was non-existent within the Garden. We bought our first motor lawnmower together with a motor lorry. He (Gilbert-Carter) although against mechanisation within the Garden, was wise enough to accept the fact that it was inevitable, and rather reluctantly asked me to dispose of the horse, Tom II, and his trappings. Tom was of an unknown quality, being only semi-trained, and it is said that when he was put out on the Fen at weekends the staff spent most of Monday trying to round him up; consequently he was reduced to spending his rest hours confined to his stable, where Humphrey delighted in feeding him handfuls of herbage, with *Symphytum* as a treat. Such was his sympathetic feeling for all animals.

The happy, successful partnership of John Gilmour as Director and Bob Younger as Superintendent in the task of extending the Garden to its total site is such recent history that the process they so effectively initiated and shaped is not yet quite complete. So far as University teaching and research are concerned, the combination of Cory money and new direction brought about within ten years the realisation of the vision that Bateson had had half a century earlier, namely the establishment on the former allotment land of the present 'Research and Experimental Area', including a new laboratory and service building erected in 1956 and its adjacent glasshouse and field plot facilities. Within the Garden itself, the new Director instituted a two-year student gardener training scheme, which happily combined the need for a larger labour force with the provision of valuable training in horticultural botany. For the general public, the 'new look' was most obvious in the construction of the limestone rock-garden by the Lake, and in the entirely new features in the 'New Area', such as the scented garden and the unique chronological bed.

Figure 81   View of the Garden in 1953, showing in the background loads of limestone ready for the new rock garden.

Figure 82   Staff of the Garden in 1959, at the height of the post-war expansion using the Cory Fund.

In deciding the use of the Cory Bequest, the Managers found an early difficulty. The terms of the will were quite restrictive, most of the very considerable sum of money being available only for capital development of the Garden. The Managers sensibly sought legal advice on the interpretation of this restrictive clause, which was then varied so as to make possible the use of the money for recurrent management expenditure on any feature in the Garden established with Cory capital. This remains the position today; since the whole of the 'New Area' and its facilities are features made with Cory capital, the restriction is no longer a serious hindrance to planning, and nearly one-third of the present Garden staff are paid from the Cory Fund. Increasingly through the 1960s and 1970s it became apparent that the division of financial responsibility should move in the direction of Cory finance being available for features of horticultural and amenity merit, whilst the University recognised more formally its direct responsibility to pay for those areas of the Garden strictly necessary for its teaching and research.

Whole-plant botany: new developments

Tansley's strong plea, with which we ended the previous chapter, was for a botany course which combined scientifically both form and function, and resisted the temptations of narrow specialisations too early in a botanical career. This was not an easy position to defend through the inter-war years, which saw the development of a highly effective and powerful Sub-department of Mycology and Plant Pathology on the one hand, and of Plant Physiology on the other; both received great impetus from a large grant from the Rockefeller Foundation with which a new wing was added to the Botany School, and a new Field Station developed in Storey's Way. Nevertheless, the position was held, and the course remained one of reasonable balance – here I can for the first time speak from personal experience! – perhaps because all specialisms were being pursued under one roof, and staff and research students alike were helpfully accessible to the keen but shy undergraduate who 'liked plants'. It was no accident that both members of staff who supervised and influenced me as an undergraduate and later as a research student after the Second World War had themselves done their own research in plant physiology. Both presented in their different ways that vision of botany which was Tansley's special concern, and my research supervisor, Sir Harry Godwin, was later able as Professor of Botany to put into practice his breadth of genuine interest in the subject as a whole with such outstanding effect.

A special feature of Cambridge field botany in the thirties was the growth of interest in bryology which began when P. W. Richards

came up to Trinity College in 1927 and began his field collecting in
the Cambridge area; these excursions became a more recognised
part of the teaching when, with Brooks' approval, he began to run
Departmental excursions for bryophytes in 1936, so beginning a
tradition which is strongly active today. Indeed, some would say
that a special mark of the Cambridge field botanist is that he knows
his mosses and liverworts as well as his vascular plants, and may
even extend his botanising to cover lichens and at least the higher
fungi as well.

The 1950s saw the rise of molecular and cell biology, a spectacular advance which has understandably dominated biological
teaching and much of biological research for the last two decades. It
is no part of my theme to assess the history or direction of cell
biology as a discipline or group of disciplines, but its impact on
botany, and on biology as a whole, is obviously relevant. At one,
rather practical, level of analysis this impact is easy to assess – it
produced a formal division between 'cells' and 'organisms', and an
abandonment of the traditional division, at least in Part I of the
Natural Sciences Tripos, between botany and zoology. As with all
such rearrangements to accommodate scientific advance it is easy to
see in them both advantages and disadvantages. The clear advantage is that the traditional division between the physical and the
biological sciences is obscured and eroded, with obvious gains on
both sides, and new disciplines like biophysics can arise and develop
in this atmosphere. The disadvantage is that inherent in any large
injection of new material, namely that, to make it possible, traditional areas of the subject are abandoned in a somewhat haphazard
manner.

Figure 83   The limestone mound in
the Ecological Area. All the plants are
of known wild origin in the British
Isles.

There is another way to look at the recent changes in the teaching
of botany which is both broader and more directly relevant to our
theme. Both the rise of ecology, involving study of the organism in
the environment, and the rise of cytogenetics, merged later into cell
biology, involve abandoning the fixed separation between botany
and zoology. We may recall that Bateson's Chair was created in
1908 as a Chair of *Biology*, and the early geneticists were rightly
impressed by the integrative effect of the new science. The movement towards teaching biology is a strong one, supported by recent
developments on both sides of the divide between 'cells' and
'organisms', and we must expect and adapt to new groupings and
new Departments, such as that of Applied Biology, the successor to
the Department of Agriculture in Cambridge.

The modern, integrative 'organismal' biology increasingly pre-
sents itself to the student, if not yet to all his teachers, as being to
some degree ecological. It is, of course, easy to decry the popular-
isation of the technical term 'ecology' in the movement of concern

for the environment which we have seen growing through the 1970s, but it is nevertheless true that projects of ecological research now, for example, make up the main part of the total research use of the Experimental Area of the Garden in 1980, whilst in 1950, of only five projects reported under 'Research' in the Annual Report, not one is said to be ecological. (To be fair, this is partly a matter of definition, for the research stocks of *Thymus* grown by C. D. Pigott and reported in that year, though described as 'for cytogenetical investigation' and 'for experimental taxonomic study', might equally well have been called 'autecological'.) The value of an ecological approach was always visible also in that school of plant physiology concerned with the behaviour of the whole plant, which saw its beginnings in the classic studies of F. Kidd, C. West and G. E. Briggs on the growth of sunflowers (*Helianthus annuus*) in 1920, and which was powerfully developed by G. C. Evans, pupil of Briggs, in recent years. Evans' role during the Professorship of Briggs (1947–60) in planning the development of the research facilities in the Garden, and in moving back into the research area the plant physiological studies carried on in the Storey's Way Field Station, have been especially important in strengthening this integration; in his book *The Quantitative Analysis of Plant Growth*, published in 1972, he uses an impressive series of data derived from the growth of sunflowers and compiled by successive Part II Long Vacation classes based in the Botanic Garden since the establishment of the Research Area.

Part of the popular interest in ecology which is so obvious in our society today is concerned with the conservation of nature and its opposite, pollution and destruction of natural communities. Inevitably this concern moves administrators, politicians, academics and students in different ways and to different effect, but, however we measure it, the revolution in public awareness and discussion of problems of the environment is a very remarkable feature of the past decade. Its impact on Cambridge botany, as on biology in general, is strongly in favour of an ecological approach, and its influence on policy in Botany School and Garden alike is unmistakable. In the examinations for natural science, nature conservation may remain an 'essay' or 'discussion' topic, but its influence on research and career possibilities is already obvious. We have housed the office of CAMBIENT, the County Naturalists' Trust, in the Garden since its birth in 1955, and more recently we have established, under contract to the Government Nature Conservancy Council, a section devoted to monitoring and cultivating the nationally rare plants of Eastern England, a region for which Cambridge provides a natural centre. This section is developing, with the help of a grant from the Pharmaceuticals Division of Imperial

Figure 84   Conservation display in the Alpine House, 1979.

Chemical Industries, given in return for samples of British plants, into a British floristic and ecological section in its own right. It is in a direct line of tradition with the Old Walkerian Garden in which, nearly two centuries ago, James Donn had in cultivation a quite exceptionally rich collection of British plants.

One other influential line in post-war botany in Cambridge is associated with the name of E. J. H. Corner, now Emeritus Professor of Tropical Botany, whose international reputation is based on expert knowledge in mycology, angiosperm taxonomy and tropical botany, communicated in a very stimulating and enthusiastic way. His former students are now in positions of responsibility in tropical institutes, and in the principal herbaria of the world where the main taxonomic research is still done on tropical plants. This interest in tropical botany seems likely to continue, at least in a broad ecological context, partly because of the urgency and importance of conservation problems in tropical regions.

In these, as in several other ways, we can now see the Garden fulfilling a role not essentially different from the one which Walker and Henslow in their different periods foresaw, perhaps more consciously and, so far as conservation is concerned, more urgently and practically. The new concern for an ecologically-based conservation role is particularly fitting in the Tansley tradition, for the whole of the official provision for the conservation of nature in this country is heavily indebted to Tansley, who typically saw the need and offered both an outline scheme and his personal service in the post-war establishment of the Nature Conservancy. As a representative of that generation – the first – which grew up in an atmosphere of concern for the fate of wild plants and animals, I can see my personal indebtedness to Tansley through the influence of his pupil, Sir Harry Godwin. A Cambridge tradition of enthusiasm for the whole plant 'in the field', involving scientific study in the garden and in the laboratory, is something which I have come to appreciate with increasing force over the years: it is a tradition which shows every sign of helping future generations as it undoubtedly helped past ones.

Bibliography

Albu, K. M. (1956), John and Thomas Martyn; a bibliography. Unpublished thesis for University of London Diploma in Librarianship. (Copy in Cory Library.)

Allen, D. E. (1967). John Martyn's Botanical Society: a Biographical Analysis of the Membership. *Proc. Bot. Soc. Brit. Is.* 6, 305–24.

Allen, D. E. (1976). *The Naturalist in Britain*. London.

Anon. (? Salton, P.) (1794). *Horti Botanici Cantabrigiensis Catalogus*. Cambridge.

Anon. (1833). Note in *Loudon Mag. Nat. Hist.* 6, 397.

Anon. (1898). Botanic Gardens of the World: The Cambridge Botanic Garden. *Pharm. Jour.* 372–4.

Arber, A. (1912; 2nd edn, 1938). *Herbals: a Chapter in the History of Botany*. Cambridge.

Babington, C. C. (1834). *Flora Bathoniensis*. Bath.

Babington, C. C. (1843). *Manual of British Botany*. London.

Babington, C. C. (1860). *Flora of Cambridgeshire*. London.

Babington, C. C. (1897). *Memoirs, Journal & Botanical Correspondence of C. C. Babington*. Cambridge.

Boutilier, J. A. (1975). 'Mr. Maudslay': Victorian Exemplar. *History Today* 25, 680–8.

Bower, F. O. (1938). *Sixty Years of Botany in Britain*. London.

Bradley, R. (1715?). *A short historical account of Coffee*. London.

Bradley, R. (1716–27). *Historia Plantarum Succulentarum*. London.

Bradley, R. (1717–18). *New Improvements of Planting and Gardening*. London.

Bradley, R. (1721). *A Philosophical Account of the Works of Nature*. London.

Bradley, R. (1721). *The Virtue and Use of Coffee*. London.

Bradley, R. (1721–4). *A General Treatise of Husbandry and Gardening*. London.

Bradley, R. (1725). *A Survey of the Ancient Husbandry and Gardening*. London.

Bradley, R. (1727). *The Country Gentleman and Farmer's Monthly Director*. London.

Bradley, R. (1728). *Dictionarium Botanicum*. London.

Bradley, R. (1729). *The Riches of a Hop-garden*. London.

Bradley, R. (1730). *A Course of Lectures on the Materia Medica*. London.

Bradley, R. (1964; facsimile, with Introduction by G. Rowley). *Collected Writings on Succulent Plants*. London.

Clapham, A. R., Tutin, T. G. & Warburg, E. F. (1952). *Flora of the British Isles*. Cambridge.

Clark, J. W. (1904). *A Concise Guide to the Town and University of Cambridge*. Cambridge.
Clark-Kennedy, A. E. (1929). *Stephen Hales, D.D., F.R.S.* Cambridge.

Darwin, C. (1888). *Autobiography* in Darwin, F. (ed.) *Life and Letters of Charles Darwin* (3 volumes). London.
De Candolle, A. P. (1832). *Physiologie Végétale*. Paris.
Donn, J. (1796). *Hortus Cantabrigiensis*. Cambridge. (For details of 13 edns up to 1845, see Henrey, B. 1975.)

Elwes, H. J. & Henry, A. (1906–13). *The Trees of Great Britain and Ireland* (7 volumes). Edinburgh.
Evans, G. C. (1972). *The Quantitative Analysis of Plant Growth*. Oxford.
Ewen, A. H. & Prime, C. T. (1975). *Ray's Flora of Cambridgeshire*. Hitchin.

Fletcher, H. R. & Brown, W. H. (1970). *The Royal Botanic Garden Edinburgh*. Edinburgh.
Freeman, J. (1852). *Life of the Rev. William Kirby, M.A.* London.

Gardiner, W. (1890–1). *Syllabus of a Course of Practical Botany for Use at the Botanical Laboratory, Cambridge*. Cambridge.
Gerard, J. (1597). *The Herball or General Historie of Plantes*. London.
Gilbert-Carter, H. (1913) *Genera of British Plants*. Cambridge.
Gilbert-Carter, H. (1922; 2nd edn, 1947). *Guide to the University Botanic Garden Cambridge*. Cambridge.
Gilbert-Carter, H. (1930; 2nd edn, 1932). *Our Catkin-bearing Plants*. Oxford.
Gilbert-Carter, H. (1934). Translations from the Danish of C. Raunkiaer, *The Life Form of Plants and Statistical Plant Geography* (collected papers). Oxford.
Gilbert-Carter, H. (1936). *British Trees and Shrubs*. Oxford.
Gilbert-Carter, H. (1937). Translation from the Danish of C. Raunkiaer, *Plant Life Forms*. Oxford.
Gilmour, J. S. L. (1944). *British Botanists*. London.
Gilmour, J. S. L. (ed.) (1972). *Thomas Johnson: botanical journeys in Kent and Hampstead* (facsimile reprint and translation). Pittsburgh, U.S.A.
Gilmour, J. S. L. & Walters, S. M. (eds.) (1975). *Humphrey Gilbert-Carter: a Memorial Volume*. Cambridge.
Godwin, H. (1977). Sir Arthur Tansley: the Man and the Subject (The Tansley Lecture 1976) *Jour. Ecol.* 65, 1–26.
Gorham, G. C. (1830). *Memoirs of John Martyn . . . and of Thomas Martyn . . .* London.
Green, J. R. (1914). *A History of Botany in the United Kingdom*. London.
Green, V. H. H. (1969). *The Universities*. London.
Grew, N. (1682). *The Anatomy of Plants*. London.

Hales, S. (1727). *Vegetable Staticks*. London.
Henrey, B. (1975). *British Botanical and Horticultural Literature Before 1800* (3 volumes). Oxford.

Henry, A. (1910). On Elm-seedlings showing Mendelian Results. *Jour. Linn. Soc. Bot. 39*, 290–300.

Henslow, J. S. (1821). Supplementary Observations to Dr Berger's Account of the Isle of Man. *Trans. Geol. Soc. 5*, 482–505.

Henslow, J. S. (1821). Geological Description of Anglesea. *Trans. Camb. Phil. Soc. 1*, 359–452.

Henslow, J. S. (1831). On the Examination of a Hybrid *Digitalis*. *Trans. Camb. Phil. Soc. 4*, 257–78 (see also Walters, S. M. ed. 1981).

Henslow, J. S. (1833). *Sketch of a Course of Lectures in Botany* (pamphlet). Cambridge.

Henslow, J. S. (1834). *Address to the Reformers of the Town of Cambridge* (pamphlet). Cambridge.

Henslow, J. S. (1836). *The Principles of Descriptive and Physiological Botany*. London.

Henslow, J. S. (1846). *Address to the Members of the University of Cambridge on . . . the Botanic Garden* (pamphlet). Cambridge.

Henslow, J. S. (1851). *Questions on the Subject-matter of Sixteen Lectures in Botany* (pamphlet). Cambridge.

Henslow, J. S. & Lamb, J. (1826). *Remarks on the Payment of the Expenses of Out-voters at an University Election* (pamphlet). Cambridge.

Hickey, M. & King, C. (1981). *100 Families of Flowering Plants*. Cambridge.

Hort, A. (1916). *Theophrastus: Enquiry into Plants* (2 volumes). London.

Hudson, W. (1762). *Flora Anglica*. London.

Hurst, C. C. (1925). *Experiments in genetics*. Cambridge.

Jenyns, L. (1862). *Memoir of the Rev. John Stevens Henslow*. London.

Johnson, T. (1629). *Iter Plantarum . . .* London.

Johnson, T. (1632). *Descriptio Itineris Plantarum . . .* London.

Johnson, T. (ed.) (1633). *The Herball or General Historie of Plantes* (Gerard). London.

Keynes, G. (1951). *John Ray: a bibliography*. London.

Kidd, F., West, C. & Briggs, G. E. (1921). A quantitative analysis of the growth of *Helianthus annuus*. *Proc. Roy. Soc. B, 92*, 368–84.

Lindley, J. (1832). *Introduction to Botany*. London.

Linnaeus, C. (1737). *Flora Lapponica*. Amsterdam.

Linnaeus, C. (1737). *Genera Plantarum*. Leiden.

Linnaeus, C. (1737). *Critica Botanica*. Leiden.

Linnaeus, C. (1751). *Philosophia Botanica*. Stockholm.

Linnaeus, C. (1753). *Species Plantarum*. Stockholm.

Lynch, R. I. (1897). *List of Ferns and Fern Allies . . . cultivated in the University Botanic Gardens* (sic) *Cambridge* (pamphlet). Cambridge.

Lynch, R. I. (1900). Hybrid Cinerarias. *Jour. Roy. Hort. Soc. 24*, 269–74.

Lynch, R. I. (1900). The Evolution of Plants illustrated by the Cultivated Nature of Gardens. *Jour. Roy. Hort. Soc. 25*, 1–21.

Lynch, R. I. (1904). *The Book of the Iris*. London.

Lynch, R. I. (1905). Gerbera. *Flora and Sylva 3*, 206–8.

Lynch, R. I. (1915). Trees of the Cambridge Botanic Garden. *Jour. Roy. Hort. Soc. 41*, 1–20.

Lynch, R. I. (1924). *Platanus digitata* Gordon, & *Platanus cantabrigiensis* Henry. *Gard. Chron. 76*, 250–1.

Lyons, I. (1763). *Fasciculus plantarum circa Cantabrigiam nascentium.* London.

Martyn, J. (1726). *Tabulae synopticae plantarum officinalium.* London.

Martyn, J. (1727). *Methodus plantarum circa Cantabrigiam nascentium.* London.

Martyn, J. (1728–37). *Historia plantarum rariorum.* London.

Martyn, J. (1729). *The First Lecture of a Course of Botany.* London.

Martyn, J. (1732). *Translation of Tournefort's History of Plants growing about Paris, with additions; accommodated to Plants growing in Great Britain.* (2 volumes) London.

Martyn, J. (1741). *The Georgics of Virgil, with an English Translation, and Notes.* London.

Martyn, J. (1749). *The Bucolics of Virgil, with an English Translation, and Notes.* London.

Martyn, J. (1754). Remarks concerning the sex of Holly. *Phil. Trans. Roy. Soc. 48*, 613–16.

Martyn, T. (1763). *Plantae cantabrigienses . . .* London.

Martyn, T. (1764). *Heads of a Course of Lectures on Botany* (pamphlet). Cambridge.

Martyn, T. (ed.) (1770). *Dissertations and Critical Remarks upon the Aeneids of Virgil . . . by the late John Martyn.* London.

Martyn, T. (1771). *Catalogus Horti botanici Cantabrigiensis.* Cambridge.

Martyn, T. (1772). *Mantissa plantarum Horti botanici Cantabrigiensis.* Cambridge.

Martyn, T. (1785). *Letters on the Elements of Botany, translated from Rousseau; with 24 additional Letters.* London.

Martyn, T. (1788). *Thirty-eight Plates with Explanations.* London.

Martyn, T. (1793) (ed. 2, 1796; ed. 3, 1807) *The Language of Botany . . .* London.

Martyn, T. (ed.) (1807). *The Gardeners' and Botanists' Dictionary; by the late Philip Miller F.R.S.* (2 volumes). London.

Monk, J. H. (1830; 2nd edn, 1833). *The Life of R. Bentley.* London.

Moss, C. E. (1914–20). *Cambridge British Flora.* Cambridge.

Mudd, W. (1861). *Manual of British Lichens.* London.

Parkinson, J. (1629). *Paradisi in sole Paradisus terrestris.* London.

Parkinson, J. (1640). *Theatrum Botanicum.* London.

Prantl, K. A. E. (trans. S. Vines) (1880). *Elementary Botany.* London.

Preston, F. G. (1940). University Botanic Garden, Cambridge *Jour. Roy. Hort. Soc. 65*, 171–81.

Raven, C. E. (1942; 2nd edn, 1950). *John Ray, Naturalist: his life and works.* Cambridge.

Raven, C. E. (1947). *English Naturalists from Neckam to Ray.* Cambridge.

Ray, J. (1660). *Catalogus Plantarum circa Cantabrigiam nascentium.* Cambridge.

Ray, J. (1670). *Catalogus Plantarum Angliae.* London.

Ray, J. (1682). *Methodus Plantarum . . .* London.

Ray, J. (1688–1704). *Historia Plantarum* (3 volumes). London.

Ray, J. (1691). *The Wisdom of God manifested in the Works of Creation.* London.

Relhan, R. (1785). *Flora Cantabrigiensis.* Cambridge. (For details of supplements and later editions, see Henrey, B. 1975.)

Roberts, W. (1939). R. Bradley, Pioneer Garden Journalist. *Jour. Roy. Hort. Soc.* *64*, 164–74.

Rowley, G. (1954). Richard Bradley and his 'History of Succulent Plants' 1716–1727. *Cact. Succ. Jour. Gt. Brit.* *16*, 30–31, 54–5, 78–81.

Rowley, G. (1964). Introduction to facsimile of *Richard Bradley's Collected Writings on Succulent Plants.* London.

Russell-Gebbett, J. (1977). *Henslow of Hitcham.* Lavenham.

Seward, A. E. (1931). *Plant Life Through the Ages.* Cambridge.

Sibthorp, J. (1794). *Flora Oxoniensis.* Oxford.

Smith, J. E. (1818). *Considerations respecting Cambridge, more particularly relating to its Botanical Professorship* (pamphlet). London.

Smith, J. J. (1840). *The Cambridge Portfolio.* Cambridge.

Tansley, A. G. (1917). On Competition between *Galium saxatile* L. (*G. hercynicum* Weig.) and *Galium sylvestre* Poll. (*G. asperum* Schreb.) on different types of soil. *Jour. Ecol.* *5*, 173–9.

Tansley, A. G. (1924). Some Aspects of the Present Position of Botany. (Presidential Address, Section K.) *Rep. Brit. Ass. Adv. Sci. 1923*, 240–260.

Thomas, H. H. (1937). The Rise of Natural Science in Cambridge. *Cambridge Review*, 434–6.

Thomas, H. H. (1952). Richard Bradley, an Early Eighteenth Century Biologist. *Bull. Brit. Soc. Hist. Sci.* *1*, 176–8.

Tjaden, W. (1973–6). Richard Bradley, F.R.S. 1688–1732, Succulent Plant Pioneer. *Bull. Afr. Succ. Pl. Soc.* *8–11* (18 papers in these 4 volumes).

Turner, W. (1551). *A new Herball.* London.

Tutin, T. G. *et al.* (eds.). (1962–80). *Flora Europaea* (5 vols). Cambridge.

Walker, R. (1762). *A short account of the . . . donation of a Botanic Garden to the University of Cambridge . . .* Cambridge.

Walters, S. M. (1961). The Shaping of Angiosperm Taxonomy. *New Phytol.* *60*, 74–84.

Walters, S. M. (ed.) (1981). Facsimile of *Henslow's paper 'On the Examination of a hybrid Digitalis'.* Cambridge.

Ward, H. M. (1891). A Model City; or, Reformed London. *New Review 5*, 182–92.

Ward, H. M. (1901), *Grasses.* Cambridge.

Ward, H. M. (1904–9). *Trees* (4 volumes). Cambridge.

Williamson, R. (1955). The germ theory of disease: neglected precursors of Louis Pasteur (Richard Bradley, Benjamin Marten and Jean-Baptiste Goiffon). *Annals of Science 2* (1), 44–57.
Winstanley, D. A. (1935). *Unreformed Cambridge*. Cambridge.
Winstanley, D. A. (1940). *Early Victorian Cambridge*. Cambridge.

Index

Page-numbers in italics refer to the captions of illustrations; normally there is also a text reference on the same page.

Acanthopanax wardii 86
Ackermann, R. *44*
Act of Uniformity 6
Adamson 92
agricultural botany 93
Agriculture Department 107
Albert, Prince Consort 67
Albu, K. M. 40
Aldrovanda 79
Allen, D. E. 30, 65
alpine plants 46
Amsterdam 19
    Physic Garden 19, 21, 25
Ananas 17
Anatomy School 30
Anglesey *50*
Annan, N. 65
Antiquarian Society 68
Apothecaries, Society of 30
Applied Biology Department 107
Arber, A. 1
Arboretum 60, 62, 72, 73, 89
Arnold, R. 34
Aristotle 4
Austin Friars 43

Babington, C. C. 4, 58, 61, 62, 65–70, 75, 76, 77, 83
    herbarium 70, 80
Baillie Scott 98
Balfour, Professor 58
Balle, R. 17
Barclay family 65
Bateson, W. 81, 90, 91, 104, 107
Bentley, Dr, Master of Trinity College, Cambridge 23
Bennett, A, 82
Berberis 102
Biffen, R. H. 93
Biggs, A. 46

'Biology of Organisms' 84
Blackman, F. F. 86
Bobart, J. 10, 11
Botanic Garden, Cambridge 53–62, *et passim*
    Act for purchase of land 55, 56, 59
    aerial photograph *102*
    Alpine House *108*
    Annual report 78, 79, 80, 82, 90, 95, 98, 102
    Chronological Bed *104*
    Curator 72, *74*, 75, *78*, 82
    Director 95, 96, 98, 103
    Ecological Area and Beds 93, *107*
    entry books 78
    Experimental Area 107
    'Friends of the Botanic Garden' 100, 103
    gates 44, *83*
    glasshouse range 77, 78, *79*, *101*
    lake 62, 73
    'New area' 100, 104, 105
    Research Area 103, 104, 107
    Research laboratory 77
    Rock Garden, new 104, *105*
    seed list 78
    'Somerset Employment Fund' 103
    Special Appeal (1932) 100
    Sunday Keys 76, 100
    Sunday opening 76
    Superintendent 75, 95, 98, *100*
    Syndicate 60, 61, 72–5, 77, 87, 90, 92, 95–9, 102
    Terrace Garden 76, *79, 92*
Botanic Gardens,
    Cambridge *passim*
    Kew 26
    Oxford 1, *10*, 11, 43

Botanic Gardens – *contd.*
    Padua 1
    Pisa 1
    St Petersburg 78
Botanical Museum 53
Botanical Society 30
Botany School, Cambridge 1, 23, 32,
    35, 82, 83–94, *86*
Boutilier, J. A. 74
Bower, F. O. 71, 89, 91
Boyd of Paisley 77
Bradley, R. 15–30, 42, 66, 85, 93
Briggs, G. E. 107
British Association 63, 69, 86, 88, 94
British Ecological Society 91
British Museum 8, 48
Brooklands Avenue *56*
Brooks, F. T. 84, 91, 103, 106
Browne, W. 10
Brownell, Mr of Willingham 26, 27
bryology 106
Brydges, J. 21
Buddleia 79, 80
Bullen, G. 60
Burghley, Lord 9
Buxton family 65
Buxus (box) *54*, 93

cacti 18, *20*
Cadbury family 65
Caesalpinus 31
'Cambridge Catalogue' 6, 8, 9, 69
Cambridge Colleges,
    Bennet 26
    Christ's 70, 84
    Corpus Christi 52
    Downing 62, 87
    Emmanuel 26, *34*, 35, 37, 87
    King's 40
    Queens' 24
    St John's 30, 41, 48, *65*, 67, *69*, 87
    Sydney Sussex 37
    Trinity 23, 71, 91, 106
    Trinity Hall 54, 55
Cambridge Natural History Society
    68
Cambridge Philosophical Society 50
Cambridge University,
    Lecturers and Demonstrators,
    Botany 80, 86, 87, 91, 96, 99
    Forestry 89
    Professors and Readers (incl.
    Chairs and Departments)
        Agriculture 93, 107

Applied Biology 107
Biology 90, 107
Botany 3, 4, 16, 18, 21, 27,
    34, 36, 47, 49, 52, 53, 58,
    70, 80, 85, 86, 90, 91, 92,
    103, 107
Forestry 89
Genetics 90
Geology 48
Mineralogy 47–50
Zoology 3, 4
Campden House, Kensington 19
Carex 45
Catharanthus roseus 45
Cedrus libani (Cedar of Lebanon) *101*
cell biology 106, 107
Chelsea Physic Garden 30, 32, 43
Cholsey-cum-Moulsford 53
Cineraria 81
'Clapham Sect' 65
Clapham, A. R. 96
Clark, J. W. 86
Clarke, E. D. 48, 49
comparative anatomy 67
coffee, *15*
        Historical account *19*
        Introduction of 26
conservation *108*
Corner, E. J. H. 108
Cory, R. 98, 99, 100
        Cory Bequest 74, 102
        Cory Cup 91
        Cory Fund 103, 105
        Cory Library 45
        Cory Lodge *96*, 98
        Cory Managers 103
County Floras 39, 40
Courtauld, W. J. 100
Cumming, Professor 48
Cytisus battandieri *102*

Darwin, C., 50–2, 64, 74
        autobiography 64
        Darwinian revolution 67
        Darwinism 63
Darwin, F. 72, 80, 86
Daubeny family 65
de Candolle, A. P. 66
Department of Botany, see Cambridge
    University
Devil's Dyke *6*, *14*
Dianthus 28
Digitalis 50
Dioscorides 4

Dipsacus strigosus 47
Donn, J.   43, 45–7, 108

Eachard, Dr   16
Eastern Counties Bus Company   99
ecology   91–3, 107, 108
    'Ecological Area' of Garden   *107*
    ecological beds   92
    ecological research   92
Economic Botanist to the Botanical
    Survey of India   95
Edinburgh, Professor of Botany and
    Materia Medica   42
Elliston, M.   37
Elodea canadensis   65
Elwes, H. J.   *89*
Engledow, F.   93
Englerian system   97
Entomological Society   68
Evans, G. C.   107

Fairchild, T.   28
Field, J.   9
floral morphology   31
folk taxonomies   3
Forestry, Chair of,
    Cambridge   89
    Dublin   89
Foster, M.   71, 77

Galium   92
garden stock   91
Gardiner, W.   80
genetics,   81
    research in Botanic Garden   90
Genetics Institute   90
Geological Society   88
Geranium sanguineum   *6, 10*
Gerard, J.   8, 9, 12
Gerbera   80
Gilbert-Carter, Sir G. T.   95
Gilbert-Carter, H.,   81, 93, 95–104
    portrait   *95*
Gilmour, J. S. L.   8, 67, 104
Ginkgo   87
Godwin, H.   91, 92, 106, 109
Gorham, G. C.   23, 37
Grew, N.   24, 31
Green, C.   31
Green, J. R.   72
Green, V. H. H.   2
Gregory   86

Hales, S.   *15*, 16, *24*

Harris, T. M.   87
Heberden, Dr   *41*
Helianthus   107
Henry, A.   89
Heffer's Printing Works   100
Henrey, B.   17
Henslow, J. S.   47–72, 84, 97, 101, 109
    Botanical Diagrams   *51*
    portrait   *64*
    village excursion   62, *63*
Herb gardens   1
Herbals   1
Herbarium, Cambridge University,
    35, 70
    Curator   93, 96
herborising   8, 39
Hermann   12
Hey, J.   37
Hickey, M.   *99*
Hill   86
Himalayan blue poppy   86
Hitcham   53
Hofmeister, W. F. B.   67
holly, dioecism of   35
Hooker, W. J. and J. D.   65
Hope, J.   42, 44
Houstoun, W.   35, 53
Hubbard, Rev H.   37
Hudson, W.   39
Hume, E. M.   92
Hunnybun, E. W.   82
Hurst, C. C.   *91*
Huxley, T. H.   63, 67, 70, 83

Ilex aquifolium, dioecism of   35
Imperial Chemical Industries   108
International Botanical Congress   88
International Congress of Botany and
    Horticulture   78
International   Phytogeographic
    Excursion   91
Iris   77, 79

Jenyns, H.   50, 65
Jenyns, L.   48, 50, 51, 52, 58, 59
John Innes Horticultural Institute   90
Johnson, T.   8

Kester, R.   100
Kew, Royal Botanic Gardens   26, 65,
    75, 79
Keynes, G.   6
Kidd, F.   107
King, C. J.   *99*

Kirby, W. 45
Kirkall, E. 32
Koch, K. H. E. 69

laboratory botany 67
laboratory, new 80
Lamb, Rev. J. 52
Lansdowne manuscripts 8
Lapidge, E. 59, 60, 61, 62
Lathraea clandestina 80
Leiden, Physic Garden 19
lichens 74
lime tree, commemorative 57, 83
Linnaeus, 12, 35, 36, 38
    Linnaean system 38, 39, 44, 66
Linnean Society 40, 69
Linum perenne 39
Liriodendron 22, 99
Little St Mary's Church 47, 53
Liveing, G. D. 68
Loggan 10
London, G. 16
Lonicera 102
Lonicera caprifolium 39
Lynch, R. I. 65, 75–82, 78, 87, 92,
    95, 97, 101

Macaulay family 65
Madagascar periwinkle 45
Magnolia 99
Maidenhair tree 87
Malpighi, M. 12, 24
Mansion House, Free School Lane
    43, 44
Manson, Dr 26
Marsh, A. S. 92
Martyn Herbarium 53
Martyn, J. 10, 16, 23, 24, 25, 27, 30–5
Martyn, T. 16, 23, 25, 26, 27, 35,
    36–45, 47, 66
Martynia 30, 34
Materia Medica 24, 30, 41
Matthiola 91
Maudslay, A. 74
Maximovicz, C. J. 78
Meconopsis 86
medicinal plants 31
Miller, C. 41, 43–5
Miller, P. 27, 32
Morison, R. 10
Mortlock, J. 44
Moss, C. E. 93, 95
Mudd, W. 73, 74, 78
Murray, A. 61, 62, 72, 73

natural classification 12
Natural Sciences Tripos 59, 75, 81, 84
Nature Conservancy 108
'New Museums Site' 44, 83
Newton, Mr 46
Newton, Sir Isaac 10, 15
Nicotiana 11

Old Botanic Garden 40–7, 42, 44,
    45, 49, 53–7, 62, 75, 83, 108
    Curator 41
    gates 44
    jackdaws in 54, 55
    lecture-rooms 49
    plan 55
    Reader on Plants 41
    Superintendent 41
Opuntia cantabrigiensis 77
Ortona Bus Company 99
Oxford,
    Botanic Garden 1, 10, 11
    Chair of Botany 16, 21, 91
    Professor of Botany 65, 72, 99
    Physic Garden 43

Padua 12
Palaeocyanus crassifolius 79
Paley, 14
Pallis, M. 92
Parkinson, J. 9, 12
Part II Botany 81
pasque flower 13
Passiflora 35
Pelargonium 34
perennial flax 39
perfoliate honeysuckle 39
Petiver, J. 19, 24
physic garden 19, 26, 43, 55
Pigott, C. D. 107
Pilularia 32
pineapple 17
Pinus sylvestris 'Moseri' 92
plant breeding 28, 93
Platanus cantabrigiensis 89
Prantl, K. A. E. 71
Preston, F. G. 95, 101
Primula 11
Prunus avium 39
Pulsatilla 13
Pulteney, R. 40, 43
Punnett, R. C. 90, 91

rare plants 108
Raven, C. E. 6–14

Ray, J.   6–14, 7, 16, 23, 31, 34, 38, 66, 96
    Ray Club   68
    Ray Society   69
Reichenbach, H. G.   69
Relhan, R.   40
Richards, P. W.   106
Richmond, T.   30
Rockefeller Foundation   106
Roses   91
Rosa 'Cantabrigiensis'   91, also Frontispiece
Rousseau, J. J.   *38*
Rowley, G.   18, 28
Royal Horticultural Society,   77, 80, 82
    Award of Merit   91
Royal Society   9, 10, 31, 32, 35, 40, 69, 88
Russell-Gebbett, J.   52

Saccharum   45
Sachs, J.   71, 84
Salix   45, 95
Salton, P.   15
Saunders, Miss   90, 91
Savage, W.   26
scala naturae   3, 5
Schultes, J.   46
Sedges   45
Sedgwick, A.   48, 50
Seward, A. C.   86, 87, *88*, 90, 92, 98, 99, 100, 101
Sibthorp, J.   40
Silene gallica   14
Silene maritima   45
Sloane, Sir H.   19, 21, 31, 32, 37
Smith, J. E.   40, 47
Sophora   84
Stratton, J.   72, 73, 74, 77
Sturm   69
succulent plants   17, 18
sugar cane   45
sunflower   107
Swaffham Prior   96
Systematic beds   44, 60, 62, 72, 73

Tansley, A. G.   91–4, 106, 109
    Memorial Lecture   91
taxonomy   66
Taxus (yew)   69
Tea Phytologist   87

Theophrastus   4
Thiselton-Dyer, W.   67, 71, 81, 83, 84
Thladiantha   78
Thomas, H. H.   16, 17, 29, 99
Thymus   107
Tilia   83
Tjaden, W.   19
Tobacco plant   11
Tournefort, J. P. de   12, 31
Tradescant, J.   11
Trevelyan family   65
Trowe, G.   22
Trumpington Road   56
Tulip-tree   22, 99
Turner, W.   8, 23
Tutin, T. G.   96

Ulmus (elm)   89

Viburnum   102
Vinca rosea   45
Vines, S.   67, 70, 72, 75, 77, *78*, 80, 83, 84
Viola   39
Vulliamy, L.   *00*

Ward, F. K.   86
Ward, H. M.   67, 70, 82–7, *86*, 89, 92, 93, 94, 97
Walker, R.   40, 41, 43, 108
Walkerian Garden, *see* Old Botanic Garden
Walkerian Reader   47, 53
Warburg, E. F.   96
Watt, Dr A. S.   89
West, C.   107
White, G.   14
Wilberforce, W.   63
Wild cherry   39
Williamson, R.   28
willows   45, 95
Willughby, F.   11
Wilmott, A. J.   69
Winstanley, D. A.   48

Xanthoceras sorbifolium   100

Younger, R. W.   104

Zoology   3, 4
Zoology Department   83

Printed in the United States
By Bookmasters